COMMITTEE ON SCIENCE AND POLICY
FOR THE COASTAL OCEAN

DONALD F. BOESCH (Co-chair), University of Maryland, Cambridge
BILIANA CICIN-SAIN (Co-chair), University of Delaware, Newark
PETER M. DOUGLAS, California Coastal Commission, San Francisco
EDWARD D. GOLDBERG, Scripps Institution of Oceanography, La Jolla,
    California
SUSAN S. HANNA, Oregon State University, Corvallis
DAVID H. KEELEY, Maine State Planning Office, Augusta
MICHAEL K. ORBACH, Duke University, Beaufort, North Carolina
JOHN M. TEAL, Woods Hole Oceanographic Institution, Massachusetts

Staff

EDWARD R. URBAN, JR., Study Director
LAVONCYÉ MALLORY, Project Assistant

Preface

Coastal areas of the United States and elsewhere face pressures from a variety of sources, both from human activities and from natural fluctuations of the environment. To confront these pressures, the concepts of ecosystem management and sustainable development have become part of national and international discussions about environmental management. Although it is not yet clear how to implement ecosystem management for the sustained use of coastal areas and their resources, one thing is certain: knowledge about coastal environmental and societal processes will be fundamental to any attempt to manage coastal environments in such a way that resources can be sustained and multiple uses accommodated.

The Ocean Studies Board (OSB) is committed to promoting the science necessary for effective coastal policy and has been active in recent years in defining important issues related to natural science in coastal areas. In response to a request from the White House Committee on Environment and Natural Resources (CENR), a committee of the OSB conducted a study to provide advice about coastal science topics related to CENR's areas of responsibility. The resulting report, *Priorities for Coastal Ecosystem Science* (NRC, 1994a), advises the government about what coastal science topics are most important for improving coastal management practices. Another OSB report, *Oceanography in the Next Decade: Building New Partnerships*, pointed out that:

> Policy decisions concerning . . . interactions of the ocean with everyday life rest upon a sound scientific understanding of the ocean. To the extent that such policy decisions are to be useful, they must be consistent with the best available information about how the system works: its physics, chemistry, geology, and biology. Both the government and the scientific community as a whole must

ensure that what is known about the ocean is made available to policy makers, that what is not known is clearly stated, and that progress in furthering our basic understanding continues. (NRC, 1992b, p. 17)

Thus, scientists have an important role and responsibility in working with policymakers to ensure that coastal environmental policies are based solidly on scientific understanding. Carrying out necessary science and using it in coastal policymaking are often difficult. This report, *Science, Policy, and the Coast: Improving Decisionmaking*, offers advice to all partners in the coastal management and policymaking process to improve the use of science in the management of our coastal waters.

WILLIAM MERRELL
Ocean Studies Board, *Chairman*

Contents

Executive Summary

Coastal ecosystems are under stress from a variety of human activities, and many have experienced widespread degradation. Laws have been enacted and regulations implemented in an attempt to stem coastal environmental damage and guide responsible development, but these control measures are not always founded on adequate scientific information. Knowledge about coastal ecosystems, including the human component, is needed to enable management of these systems in a manner that will preserve their value and restore degraded systems while allowing economic development and a high quality of life. A continuous exchange of information between scientists[1] and managers who focus on coastal areas is necessary to develop and use scientific results effectively and to address emerging environmental problems in coastal areas. This need is becoming more evident as the complexity of the relationships among the environment, resources, and the economic and social well-being of human populations is recognized fully. All stakeholders—scientists, managers, industry, the public, environmental groups, and others—should be involved in coastal policy formation and management.

The National Research Council's Ocean Studies Board (OSB) and its Committee on the Coastal Ocean believed that a study to examine the existing interactions between coastal scientists and policymakers and to recommend ways to improve these interactions could be beneficial to states, regions, and the nation. To understand the use of science in policymaking and how scientists and policymakers interact, three symposia were convened—in California, the Gulf of Maine

---

[1]Unless otherwise noted, the term "scientists" is used to refer to both natural and social scientists.

1

region, and the Gulf of Mexico region—that focused on the use of science in addressing specific regional issues. The purpose of the symposia was to formulate recommendations for improving the application of science by evaluating existing practices and past successes and failures in coastal policymaking. Furthermore, it was expected that mechanisms for using science in coastal policymaking identified in each region could be transferred to other regions and used nationally and that comparisons among regions would yield additional insights. Each regional symposium was documented with a proceedings report, each providing a wealth of information. The OSB formed a Committee on Science and Policy for the Coastal Ocean to summarize and synthesize the findings of the three symposia and to make recommendations for improving the use of science in coastal policymaking and management.

This study was based on three tenets:

1. Successful coastal environmental policies have been formulated over the past half century through efforts of scientists and/or policymakers; science and technology have played an important role in policy successes.

2. Problems related to interactions between scientists and policymakers are shared among the three regions studied and presumably across the nation.

3. Experiences with the management of coastal environmental problems can provide lessons to guide future management and policymaking.

Much was learned in the regional symposia regarding the use of science in coastal policymaking and management. It was clear that scientific information is more important in some stages of the policy process than in others. Scientists and policymakers must be aware of the differences in their cultures and reward systems and create mechanisms for interaction that account for these differences. Environmental problems should be well defined, with the proper questions being asked in a language shared by scientists and policymakers. To be helpful to policymakers, science must provide timely and credible information that is responsive to policy-relevant questions. Scientists must identify the significance of their findings and the limitations inherent in the information they provide, as well as the additional questions that are raised by their research and the potential cost of addressing those questions. Great care must be devoted to providing a structure for interactions that yields scientific advice that is objective and balanced. Adaptive management systems, in which science is a substantial part of planning, evaluating, and modifying management strategies, are gaining favor as a means to improve interactions between scientists and managers for the purpose of creating more effective environmental policy. Adaptive management has the capacity to detect, learn from, and adapt to changing circumstances and new information. Integration of management efforts also is important. Integrated management involves harmonizing policy development and implementation among coastal zone uses, interacting land and sea processes, levels of government, and scientific disciplines.

Discussions at the symposia and of the Committee on Science and Policy for the Coastal Ocean revealed three common themes: (1) coastal scientists and policymakers do not interact sufficiently to ensure that decisions and policies related to coastal areas are based adequately on science; (2) coastal policies tend to lack sufficient flexibility and most often are designed to manage single issues; and (3) compared with resources allocated individually to policy, management, and science, the allocation of available resources to apply coastal science to policymaking is suboptimal. To address these concerns, the committee recommends that agencies and legislatures at state and federal levels:

1. improve the interaction between scientists (natural and social scientists) and coastal policymakers/implementors at all levels of government,
2. employ integrated and adaptive management approaches in coastal policymaking and implementation, and
3. improve the allocation and coordination of resources to achieve effective interaction between coastal scientists and policymakers.

Specifically, the committee recommends that the National Oceanic and Atmospheric Administration, the Environmental Protection Agency, the Department of Interior, the Department of Energy, the U.S. Army Corps of Engineers, and other relevant federal agencies review the recommendations herein for application at the federal level. Federal agencies could benefit from implementing the recommendations of this report through revisions to existing agency policies, programs, and practices and in the creation of new ones.

Congress should consider the recommendations contained herein in the development of legislation affecting coastal environments and their resources. The recommendations of this report are relevant to many federal laws (see Table 1 for examples), particularly the Coastal Zone Management Act.

Recommendations herein could also provide useful guidance to state agencies and legislatures. Authorities in states and regions could benefit from an analysis of region-specific suggestions summarized in Chapter 2 and discussed in more detail in the proceedings of the regional symposia.

The recommendations in this report are directed not only at governmental agencies and elected officials but also scientists and academic institutions, industry, nongovernmental organizations, the news media, and the public. Detailed actions to implement the three recommendations listed above are discussed in Chapter 4 and are summarized in Table 4 on pp. 64-67.

The committee believes that many of its recommendations could also be applied to address coastal problems faced by other nations, because many of the same problems are experienced around the world. A major recommendation of the 1992 United Nations Conference on Environment and Development was that nations should create (or strengthen) management processes and institutions to attain sustainable development of their marine and coastal areas, and we offer this report as one step toward that goal.

1

Introduction

COASTAL ENVIRONMENTS UNDER PRESSURE

The coastal[2] areas of the United States, and indeed the world, are sites of intense human activity. A majority of the U.S. and world populations live either near the coast or along rivers that empty directly into the coastal zone. Furthermore, coastal populations are growing faster than inland populations, increasing by more than 1 percent per year in the United States (Culliton et al., 1990). Coastal population growth creates extra demands for food, waste disposal, public health, and protection from natural disasters. In the United States, changing demography is resulting in more affluence and more consumption of resources and land per individual in many coastal areas. For example, in the Chesapeake Bay watershed, the conversion of farms and forests to residential and commercial uses has outstripped the rate of population increase by a factor of 3.5 (The Year 2020 Panel, 1988). In other coastal regions, people are rapidly immigrating into urban centers, resulting in changing social and economic demands on the coastal environment.

In addition to residents of the coastal zone, many additional people flock to the coast for recreation and tourism, increasing environmental pressures from roads, commercial development, waste disposal, marinas, and other recreational facilities. The resources of coastal ecosystems are exploited for seafood and

---

[2]"Coastal" is defined herein as the zone extending seaward 200 miles from the coastline to the limit of the U.S. Exclusive Economic Zone and extending landward from the coastline to the limit of tidal influence.

energy resources to benefit the entire country. Many fish and shellfish stocks from enclosed coastal waters and the U.S. Exclusive Economic Zone are being harvested to the limits of sustainability or are being overfished (NMFS, 1993). For example, the groundfishery off the northeastern United States has essentially collapsed. The previously dominant cod and flatfish species have been replaced by dogfish and skates (Fogerty et al., 1991). Oyster production from the Chesapeake Bay is now less than 5 percent by volume of what was harvested at the turn of the century (Richkus et al., 1992).

There is a growing realization that coastal areas are also influenced by events and processes that occur far from the coastline. For example, buoyant materials accidently or purposely dumped at sea find their way to beaches (Ebbesmeyer and Ingraham, 1992). Nonindigenous marine organisms may be introduced through discharge of ballast water from transoceanic ships originating from foreign ports. Organisms, such as Asian clams introduced into San Francisco Bay (presumably via ballast water) can proliferate and have major effects on coastal ecosystems (Nichols et al., 1990). Spores of algae responsible for toxic blooms can be introduced in the same manner (Hallegraff and Bolch, 1991). These and other species are of growing concern (Carlton and Geller, 1993). In San Francisco Bay, for example, at least 255 alien species of invertebrates have established populations (Hedgpeth, 1993).

Coastal ecosystems are also affected by activities that occur far inland, through changes in the delivery of water, nutrients, and chemical contaminants from rivers and atmospheric deposition (NRC, 1994b). Large areas of such important coastal waters as the Chesapeake Bay, the northern Gulf of Mexico, Long Island Sound, Lake Erie, the North Sea, and the northern Adriatic Sea have experienced increased plankton blooms and depletion of dissolved oxygen as a result of nutrient overenrichment from both point-source (sewage discharges) and diffuse inputs (agricultural and urban runoff and atmospheric inputs) during the latter half of this century (Officer et al., 1984; Rydberg et al., 1990; Parker and O'Reilly, 1991; Rabalais et al., 1994). Finally, coastal environments are among those most susceptible to the consequences of global climate change that could affect sea level, freshwater runoff, frequency and intensity of storms, and temperature patterns.

Dealing with increases in coastal populations, resource consumption, and land development, while at the same time trying to protect healthy environments, restore degraded environments, replenish depleted fisheries, support economic development, and enhance the quality of human life, is a daunting challenge for policymakers. Coincidentally, society is demanding more efficient and less intrusive governance, which requires technically sound assessments of risks, costs, and benefits among alternative decisions as well as progress toward more local decisionmaking.

THE IMPORTANCE OF SCIENCE

Scientific information is needed to guide the wise use of coastal resources, to protect the environment, and to improve the quality of life of coastal zone residents. This need is becoming more evident as the complexity of the relationships among the environment, resources, and the economic and social well-being of human populations is fully recognized and as changes and long-term threats are discovered. Earlier this century, coastal managers and policymakers concerned themselves primarily with how people could exploit coastal areas and resources, with little recognition of the impacts of such exploitation. The impacts were relatively modest (or at least localized) until coastal land and water use and the discharge of society's wastes intensified during the last half of this century. Fishing, population growth, fertilizer and pesticide use, fossil fuel consumption, shipping, wetlands destruction, and other factors began to take their toll. Environmental legislation passed by states and the federal government in the 1960s and 1970s (e.g., the National Environmental Policy Act and the Coastal Zone Management Act) gave new importance to coastal management and policymaking and focused new attention on the need for scientific information for decision-making.

Although the mounting pressures pose new challenges for management and the science to support it, it is important to recognize that there are coastal problems that have been dealt with effectively and that science and technology have played an important role in these successes. A good example is the formulation of policies and regulations regarding the release of artificial radionuclides from energy facilities to the atmosphere and the ocean (NAS, 1957; NRC, 1971). Policies implemented in the 1950s and 1960s were successful in protecting the health of human populations. In the 1960s the impacts of halogenated hydrocarbon biocides on nontarget organisms, particularly the effects of DDT and its degradation products on fish-eating birds, brought about legal constraints to their use in pest control, which have allowed the recovery of threatened species. Science played a key role in the strategies that led to the cleanup of some coastal systems degraded by overloading with organic and nutrient wastes, including Lake Erie (IAGLR, 1991) and the Potomac River below Washington, D.C. (Jaworski, 1990).

A more recent example of how science has contributed to solving coastal environmental problems is the identification of the effects of extremely low concentrations of tributyl tin (in marine paints) on marine organisms (Goldberg, 1986). This led to the banning or severe restriction of this extremely toxic compound by several nations and U.S. states. Likewise, imposition of technology-based standards for pretreatment of industrial and municipal discharges and resource recovery efforts have resulted in reductions in the loadings of many toxic substances to U.S. coastal waters and incorporation in sediments and in tissues of marine organisms (NOAA, 1990).

Some environmental and resource problems have worsened because scientific information is inadequate to understand or anticipate them fully. Widespread eutrophication (nutrient overenrichment) of coastal waters has been documented, often only after severe effects, such as complete depletion of oxygen, were apparent (Rabalais et al., 1994). Similarly, our poor understanding of the quantitative relationships between water quality, habitat condition, and fish production has made it difficult to develop management strategies for sustainable populations that separate environmental effects from those caused by overfishing.

Perhaps even more troubling have been instances in which scientific information was available and there was scientific consensus but appropriate policies were not enacted, resulting in worsening environmental conditions, declining resources, or unnecessary costs to society. The collapse of groundfish populations off the northeastern U.S. coast resulted in part because the harvests allowed by the management process exceeded those that scientific analyses indicated were sustainable. Although controversy about the environmental effects of drilling discharges was a factor that delayed oil and gas exploration, scientific consensus indicated that these discharges posed little risk (NRC, 1983a). Requirements for secondary treatment of municipal discharges from some California communities were pursued vigorously when scientific consensus indicated that there were more effective and less expensive alternatives (NRC, 1993). These "failures" were the result of weak linkages between science and policy, exacerbated by the way scientists and policymakers deal with uncertainty, how consensus is formed, and the influence of social, economic, cultural, and political factors on decisionmaking.

Scientists have long been fascinated by the coastal ocean, its organisms, and its environmental processes. The first ocean science institutes founded at the end of the nineteenth century focused on coastal studies. Many still do. As people have colonized coastal areas, scientists and engineers have increasingly been called on to provide an understanding of natural systems to help adapt human activities to coastal processes in such a way that natural systems are preserved, resources sustained, and the flow of benefits maintained. Although social scientists began coastal studies in an organized fashion later than did natural scientists, social scientists have, since the 1970s, played an increasingly important role in understanding the human element of coastal ecosystems, which is a broadly interdisciplinary challenge.

The scientific community often views environmental policy as being influenced by public pressures more than by scientific considerations. In our democratic system of government it is appropriate for policy to be influenced significantly by the public. The public plays an important role in interpreting science and in communicating its preferences to policymakers. This factor makes public education about science important. Situations sometimes arise in which there is an apparent agreement between decisionmakers and the scientific community, but the policies fail because the affected public disagrees with the policy choice. The reasons for the disagreement may include differences in goals and objectives, the financial cost of the policy, different perceptions of risk, ineffective public education strategies, misinformation and distortion of facts by individuals or groups with vested interests in particular policy outcomes, or a simple lack of trust that scientists and policymakers have taken sufficient account of public values.

Neither the importance of science (including research, monitoring, and modeling) to the wise management of coastal environments, resources, and human effects, nor the process by which science is used, has been emphasized adequately. Concerted efforts are seldom made to foster interactions between scientists (social and natural) and policymakers (agency and legislative). Often the public is involved late in policy processes and only in response to legislative and regulatory mandates. New means must be developed to improve interactions among scientists, policymakers, and the public so that policymakers can obtain the information they need about social and natural systems, scientists can determine from policymakers what kinds of scientific questions are relevant to policy, and the public can be introduced as an integral participant in coastal management and policymaking. Improved interactions among all relevant participants are needed but are unlikely to occur until attention is focused on interaction processes.

ORIGINS OF THIS ASSESSMENT

This study resulted from the realization by members of the National Research Council's Ocean Studies Board and its Committee on the Coastal Ocean that the use of science in coastal policymaking is often less effective than is desirable and should be improved. It was anticipated that improvements could be achieved by assessing existing practices, evaluating past successes and failures in the use of science in the management of coastal areas, and discussing new means of communication among scientists, policymakers, and the public to complement existing methods that seem to be effective.

U.S. federal agencies, including those that funded this assessment—the National Oceanic and Atmospheric Administration (NOAA), the Environmental Protection Agency (EPA), and the Minerals Management Service (MMS)—have made significant investments in coastal research and management. The federal government spent $672 million on coastal science in FY1991-1993, primarily for science related to living resources, habitat conservation, and environmental quality (SUSCOS, 1993). There has also been substantial national investment in coastal ocean management activities through programs such as NOAA's Coastal Zone Management Program and EPA's National Estuary Program and through the implementation of various federal laws concerned with coastal areas (see Table 1). The national investment in the marine-oriented social sciences has been

TABLE 1 Science and National Coastal/Ocean Management Programs (from Knecht, 1995)

	Explicit Science Component in Legislation/ Procedures	Tractability of the Management Problem	Degree to Which Physical Sciences are Involved	Degree to Which Social Sciences are Involved
CZMA	No	Low	Moderate	Low
FCMA	Yes	Moderate	Moderate to high	Moderate
OCSLA	Yes	Moderate to high	Moderate	Moderate
MMPA	Yes	High	High	Low
NEP	Yes	Moderate	Moderate to high	Low

CZMA - Coastal Zone Management Act
FCMA - Fishery Conservation and Management Act
OCSLA - Outer Continental Shelf Lands Act
MMPA - Marine Mammals Protection Act
NEP - National Estuary Program

modest (primarily through NOAA's National Sea Grant College Program and to some extent as a part of programs funded by agencies such as MMS). Nonetheless, a significant body of knowledge and expertise on the human aspects of coastal ocean issues is available in the United States. The use of science in coastal management programs varies; Knecht (1995) presented his estimates of the degree to which science is used in federal programs related to coastal management (Table 1).

STRATEGY USED

Regional Symposia

The strategy of this study was to gather information about specific issues of science-policy interactions in three different coastal regions of the United States: California, the Gulf of Maine, and the Gulf of Mexico (see Figure 1). It was expected that different interactive processes identified in each region could be transferred to other regions and used nationally and that comparisons among regions would yield additional insights. The three regions differ in terms of such factors as political structures (the single-state California situation versus the multistate binational Gulf of Maine and Gulf of Mexico situations), the extent of policy experience in each region, and the degree to which the coastal physical

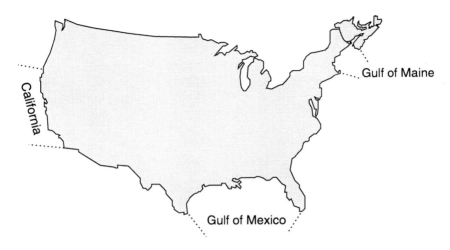

Figure 1 Regions addressed in the three regional symposia.

systems are connected throughout the region. All three regions have substantial past and present research and policy activities related to the coastal ocean.

Regional information was gathered by conducting three regional symposia that brought together natural and social scientists; legislative staff and policymakers;[3] federal, state, and local agency officials; and representatives of environmental and industrial organizations. Each symposium extended over two days and included plenary stage-setting sessions and concurrent sessions focused on specific issues. Each of the symposia addressed three types of issues: (1) one issue of common importance in all three regions, (2) one regional issue of present interest, and (3) one issue that is likely to be of importance to the region in the future (see Table 2). "Cumulative impacts of development" was selected as the common theme to be addressed in each symposium. The three themes for each region were identified by a group of individuals within the region. Discussions and interactions of participants at the three symposia, fueled by information in the proceedings reports, were intended to stimulate increased use of science in coastal policymaking in each region.

The symposia sought to:

• elucidate the process of interaction between science and policy by examining a number of case studies of successes and failures;
• identify obstacles to effective interaction between science and policy;
• identify specific incentives and mechanisms for improving the interaction between science and policy; and

---

[3]Relatively few of the legislative staff and legislators invited were able to attend the symposia.

- develop ideas for specific actions that could improve science-policy inter-actions in the region and, ultimately, the nation.

A proceedings volume has been published for each of the regional symposia, containing the papers presented and the results of group discussions related to the three issues addressed in the region (NRC, 1995a,b,c). Although each symposium was designed as a separate activity, each was also designed to contribute to the formation of national-level recommendations for improving the use of science in coastal policymaking. Most members of the Committee on Science and Policy for the Coastal Ocean participated in all three symposia and were involved in planning from the initiation of the project. The committee used the symposia proceedings in large measure as the basis for forming the recommendations presented in this report.

OBJECTIVES OF THE REPORT

The purpose of this report is to make recommendations for improving the use of science in coastal policy and management. This objective requires that communication be improved among all participants in the coastal policy process—scientists, regulators, legislators, and the public—so that more rapid progress can be achieved in making the use of U.S. coastal areas sustainable.

The report first summarizes the regional symposia by describing the regional settings and briefly summarizing the perspectives and suggestions of the participants related to the three issues addressed at each symposium (Chapter 2). Particular attention is paid to the common thread of addressing cumulative impacts. Based on the results of the regional symposia, the broader literature, and committee members' experiences, the report then discusses the challenges for effective interactions (Chapter 3). This chapter addresses the role and limitations of science, cultural differences that affect interactions, scientific advisory and review mechanisms, the integration of scientific information, the role of prediction and uncertainty, scientific agenda setting, the interactions among the multiple sectors involved in the policy process, and making coastal management more integrated and adaptive. The committee presents its findings and recommendations for improving the use of science in coastal policymaking in Chapter 4.

The recommendations of the report are directed primarily to federal and state agency staff, federal and state legislators, and natural and social scientific communities. The committee believes that many of its recommendations could also be applied beneficially by other nations, because many of the same coastal problems are experienced around the world. A major recommendation of the 1992 United Nations Conference on Environment and Development was that nations should create (or strengthen) management processes and institutions to attain sustainable development of their marine and coastal areas, and we offer this report as one step toward that goal.

2

Regional Symposia

Interactions between coastal science and policy were examined in depth at three symposia involving scientists, policymakers, managers, industry representatives, representatives of nongovernmental organizations, and private citizens, with experience in the California, Gulf of Maine, and Gulf of Mexico regions. Proceedings for each symposium have been published separately (NRC, 1995a,b,c) and are cited extensively throughout this report. This chapter presents a brief overview of the three regions, including significant environmental and resource management concerns identified by individuals from each region and a summary of the discussions of three issues selected for particular focus during the symposium.

The three issues examined in each regional symposium are listed in Table 2. In each case one issue was selected from each of the following three categories: (1) concerns that are of intense current regional interest, (2) emerging concerns for the future, and (3) concerns related to the cumulative impacts of the human use of coastal environments.

THE CALIFORNIA SYMPOSIUM

California's coastal region is characterized by spectacular beauty and remarkable natural and cultural richness and diversity. From the fog-shrouded redwoods of the north to the wide sandy beaches in the south, the more than 1,100-mile California coast supports numerous habitats and spans several biological provinces (California Coastal Commission, 1987). Most of this coast is composed of headlands with semienclosed bays, lagoons, and estuaries, which

TABLE 2   Issues for Which Science-Policy Interactions Were Evaluated in
Each of the Three Regional Symposia

Region	Issues
California Nov. 11-13, 1992, Irvine, Calif.	Cumulative Impacts of Development Coastal Ocean Habitat Mitigation Strategies Coastal Sediment and Water Quality
Gulf of Maine Nov. 2-4, 1994, Kennebunkport, Maine	Cumulative Impacts of Land and Water Activities Protecting Regionally Significant Terrestrial and Marine Habitats Using Indicators of Environmental Quality
Gulf of Mexico Jan. 25-27, 1995, New Orleans, Louisiana	Cumulative Impacts of Offshore and Coastal Oil and Gas Development Effects of Freshwater Inflow Changes Water Quality and Shellfish Production

are, with a few exceptions such as San Francisco and San Diego bays, relatively
limited in scope. The habitats in these environments have been particularly
susceptible to the effects of human activities. For example, since the 1950s,
nearly 90 percent of California's once highly productive coastal wetlands have
been destroyed or substantially altered (California Coastal Zone Conservation
Commission, 1975). Most rivers have been dammed or modified, starving
beaches of sand and estuaries of vital fresh water. Intensive urbanization, min-
ing, logging, agriculture, tourism, development of public works such as roads,
and energy and port development have forever changed the character of
California's coast. The diversion of fresh water to supply the needs of the state's
nationally important agricultural enterprise and the domestic water needs of the
urbanized south has resulted in significant changes in the San Francisco Bay
estuary and the resources it supports. The bay is important for commercial and
military maritime purposes, but with shipping has come the introduction of many
nonindigenous species that have changed the bay's ecosystem. Furthermore,
there are continuing problems of disposal of dredged material.

To the south, the large and growing population centers around Los Angeles
and San Diego have generated concerns about the effects of sewage disposal and
polluted runoff into the ocean and loss of public access and recreational opportunities. Major controversies involve the expansion of offshore oil and gas production, which began in California in the 1880s, including the effects of oil spills,
onshore support activities, and loss of aesthetic resources. Largely as a result of
these concerns, substantial areas along the coast have been designated as state
and national marine sanctuaries, which pose new management challenges for
balancing preservation and development.

The value of the California coast to the state and nation is immeasurable. As a consequence of the importance of the coast to its people, California established one of the first and most comprehensive state coastal management programs in the nation. In contrast to the Gulf of Maine and Gulf of Mexico, most of California's coastal problems are intrastate in scope and origin. Multijurisdictional issues abound, however, as California's 73 local coastal governments (58 cities and 15 counties) and over 100 special districts and state, regional, and federal governmental entities struggle to both use and protect the coast.

Science has occasionally played an important role in aspects of coastal ocean management in California. One specific case of science-policy interaction, unprecedented in its ecosystem approach, is the California Coordinated Oceanic Fisheries Investigation (CalCOFI), established in 1948 in the wake of the collapse of the California sardine fishery (Scheiber, 1995). Science has also been applied to air and water quality control programs, monitoring programs related to public health and habitat productivity, port expansion activities, selection of offshore dredge disposal sites, location of new marinas, offshore energy resources development, mitigating the impacts of electricity generating plants, and public lands management. Research on the subject of shoreline processes and hazards affecting coastal development and uses, habitat loss and restoration, a wide range of fisheries issues, and aquaculture has helped state and local officials manage these processes and activities.

As a result of the attention that the California symposium focused on the need to increase the use of science in coastal policymaking, the California legislature amended the state coastal management program's legislation to encourage science-based decisionmaking (California Public Resources Code, 1992, Section 30335.5).

Summary of California Symposium Findings

Cumulative Impacts of Development—Consensus was achieved on several points. Cumulative impacts exist and must be addressed relative to nearly every major issue faced by coastal managers (e.g., loss of coastal habitat, air and water quality, public access, visual quality). Although such impacts are real, very few people understand them, they are difficult to measure, and governance systems are not structured or inclined to deal with them. The "tyranny of small decisions"—the unintentional and adverse consequences resulting from narrowly focused actions on individual projects—is characteristic of the management of California's coastal ocean. Significant issues relating to cumulative impacts were raised in each of the major discussion sessions and became the common thread running through this symposium and subsequent symposia.

In 1990 Congress amended the Coastal Zone Management Act (CZMA) and identified cumulative impacts of coastal development as a priority need nationwide. In response, California's Coastal Commission launched a Regional Cumulative Assessment Project (ReCAP) in the Monterey Bay region to assess the cumulative impacts of development on wetlands, coastal hazards, and public access. The project is intended to result in program and policy recommendations to improve management of such impacts (California Coastal Commission, 1994). It is too early to judge the results of this effort, which moves into its critical implementation phase in 1996. Understanding and dealing with cumulative impacts offer many opportunities for, and requires, extensive interaction between the science and policy communities.

Coastal Ocean Habitat Mitigation Strategies—Vast areas of biologically rich habitats in California are economically, environmentally, and socially significant but have been substantially altered or lost because of human perturbations. Wetlands and estuarine systems and coastal embayments containing relatively shallow waters have been affected most severely. Over the past 30 years, mitigation and restoration techniques have been developed and tested in an attempt to recover some of this valuable coastal habitat.

Although science played little or no role in actions that resulted in the loss of these resources, recent decisions about mitigation and restoration have been influenced by science. At the same time, science relating to coastal habitat creation, restoration, and enhancement is still in its infancy. Accordingly, although much has been learned, until it can be demonstrated scientifically that habitats have been successfully restored, such activities should be viewed as experimental and prevention of habitat loss should still be emphasized. Given the existing state of the art, it should not be assumed that habitats lost to new development can be replaced elsewhere through creation and restoration techniques (NRC, 1992a, 1994a). Strategies to maintain the biological integrity of coastal habitats should begin with avoidance of loss, minimization of adverse impacts, and compensation for unavoidable losses. When these strategies are unavailable, creation of new habitat and substantial restoration of severely degraded habitat should be pursued. This is an area where the use of science in policymaking is achieving positive results and where continuing interactions between scientists and policymakers are essential and will be very productive. Fundamental limitations include lack of basic information about the ecosystems affected, lack of agreement on valid methodologies to accurately measure the values of the habitat lost or adversely affected by new development and the habitat to be restored, and the absence of mechanisms to make scientific information available in a timely fashion for policymaking and management.

Despite the obstacles described above, examples of meaningful and effective science-policy interactions were identified and discussed. The knowledge exists to improve future restoration and mitigation projects significantly. Several principles were identified. First, criteria should be adopted for the selection of restoration and mitigation sites. Project sites should be evaluated in the context of the larger physical and biological systems of which they are a part. Second, clear performance standards should be adopted for project development and for measuring, based on sound scientific evidence, the extent to which restoration or mitigation objectives are met (e.g., project success). Third, long-term monitoring should be provided to measure the effectiveness of maintenance and remediation and to identify lessons learned to inform future decisions and projects. An example incorporating these principles is the approach adopted by the California Coastal Commission to attempt to compensate for the adverse impacts on the marine environment of the operation of units two and three of the San Onofre Nuclear Generating Station in northern San Diego County. This compensation took the form of requirements that the utility restore the San Dieguito wetlands and create an artificial reef.

The development of an applied science of mitigation and restoration of coastal ocean habitat was encouraged. Creation of a special panel to identify the long-term information needs of decisionmakers, creation of a research agenda to secure this information, and assessment and synthesis of evolving mitigation science also were recommended.

Coastal Sediment and Water Quality—The quality of coastal waters and sediments in Southern California is of major public concern. The periodic need for beach closures and health warnings about eating contaminated fish justifies the expenditure of considerable resources annually for monitoring. Population growth and development have increased levels of pollution discharged into marine waters. To avoid enormous prevention and cleanup costs, better cause-and-effect data are critical, especially to distinguish between natural and human-caused impacts on biological health and productivity. Remedial management decisions have been influenced by good science, and improvements have been achieved.

However, problems persist. The causes, dynamics (e.g., additive effects), and geochemistry of anthropogenic pollutants in the marine environment involve complex processes, some of which are not well understood. Like habitat restoration, this is a young field of scientific knowledge. Existing monitoring data are often not available in usable form. Monitoring data need to be more readily accessible, sampling and analytical methodology should be updated and standardized (so that samples collected at different sites and times are analyzed by the same procedures), and quality control and assurance are needed. Additional research is also required, especially in the area of identifying, understanding, ranking, and prioritizing risks. Strategies to protect and enhance the productivity of the Southern California marine environment should be comprehensive and should integrate monitoring with research on a scale that takes into account the spatial and temporal parameters of nearshore processes.

Although the components of a comprehensive integrative approach are known, implementation is challenging. An example of this approach is the Environmental Protection Agency's National Estuary Program (NEP). The two

NEP projects in California (for San Francisco and Santa Monica bays) were among several examples examined by symposium participants to illustrate ways in which science-policy interactions operate relative to coastal water quality issues. The major unknown factor relative to these projects is the degree to which the estuary management plans prepared through these programs will actually be implemented.

THE GULF OF MAINE SYMPOSIUM

The Gulf of Maine extends from Cape Sable, Nova Scotia, to Cape Cod, Massachusetts, and includes the Bay of Fundy and Georges Bank. It is a semienclosed sea, separated from the Atlantic Ocean by underwater banks. It is an economic resource linking three American states and two Canadian provinces and is the foundation of a distinct maritime culture shared by two countries. More importantly, the Gulf of Maine is a marine ecosystem comprised of nutrient cycles, currents and tides, food chains, and energy flows. The crustaceans, fish, marine mammals, and birds inhabiting the Gulf of Maine region lead transboundary lives, crossing between Canada and the United States freely.

The foundation for the Gulf of Maine symposium was the initiative of the region's governors (Maine, Massachusetts, and New Hampshire) and premiers (New Brunswick and Nova Scotia) as expressed in the 1989 Agreement on Conservation of the Marine Environment of the Gulf of Maine Between the Governments of the Bordering States and Provinces. Through this agreement the region's leaders formed the multilateral Gulf of Maine Program, established the Gulf of Maine Council on the Marine Environment (hereafter referred to as the council), and called for the negotiation of a 10-year Natural Resources Action Plan.

Fundamental to the Gulf of Maine Program was the definition of shared environmental management goals and the desire to ". . . maintain and enhance marine environmental quality in the Gulf of Maine and to allow for sustainable resource use by existing and future generations."[4]

The Gulf of Maine Program builds on existing state, provincial, and federal initiatives. Examples of these initiatives are the three state coastal management programs and Sea Grant College Programs, the National Estuarine Research Reserves, and the National Estuary Program sites. The Gulf of Maine Program serves as a cooperative mechanism to address transboundary issues by adopting a watershed management approach that incorporates both the upland area and the coastal marine environment. It seeks to accomplish this through implementation of the Natural Resources Action Plan and a gulf-wide monitoring plan.

---

[4]Agreement on Conservation of the Marine Environment of the Gulf of Maine Between the Governments of the Bordering States and Provinces (signed by the governors and premiers of Maine, Massachusetts, New Hampshire, New Brunswick, and Nova Scotia in Portland, Maine, in 1989).

The integration of science, policy formulation, and management is facilitated by the existence of the Regional Association for Research on the Gulf of Maine (RARGOM)[5] and the Gulf of Maine Regional Marine Research Board (RMRB).[6] These organizations are actively involved in gulf-wide issues and have negotiated a three-party agreement with the council on ways to sustain cooperative activities. This coordination has been supplemented by the formation of the Gulf of Maine Collaboration of Community Foundations (CCF).[7] In 1991 the council prepared the Natural Resources Action Plan, which was subsequently adopted by the governors and premiers and represents a regional consensus on the most pressing natural resource management issues to be addressed by the council in its first 10 years.

The plan identifies specific annual actions that the council will pursue pertaining to coastal pollution, monitoring and research, public education, habitat protection, and public health. Because the Gulf of Maine Program is a state-provincial initiative, implementation of its action plan has been largely dependent on the availability of state-provincial financial resources. As the region's economy slowed in the 1990s, so did implementation of the plan. The Gulf of Maine Symposium focused on three issues that emanated from the plan. These included (1) cumulative impacts of land and water activities, (2) protecting regionally significant terrestrial and marine habitats, and (3) using indicators of environmental quality.

Summary of Gulf of Maine Symposium Findings

Cumulative Impacts of Land and Water Activities—One of the most pressing coastal and ocean management issues is the gradual, incremental degradation and loss of resources. This issue was identified by the three states in their CZMA Section 309 strategies[8] and by the council as it developed its action plan. Primary issues of concern involve the loss of coastal and marine habitats and the consequences of that loss. For example, the collapse of the groundfishery is largely the

---

[5]RARGOM is an association designed to foster quality scientific research related to the Gulf of Maine through increased communication and collaboration among the region's academic and management institutions.

[6]The nine regional marine research programs were established in 1990 by the U.S. Congress by an amendment to the Marine Protection, Research, and Sanctuaries Act.

[7]CCF was formed in 1992 to facilitate cross-border approaches to protect the ecological integrity and economic sustainability of the Gulf of Maine ecosystem and to encourage multisectoral discussions around gulf issues. It is comprised of the six individual community foundations bordering the Gulf of Maine.

[8]Section 309 of the CZMA requires states to prepare and submit to the Ocean and Coastal Resources Management office coastal assessments and strategies that respond to eight national priorities.

result of overfishing. However, there is growing evidence that the ecosystem effects of harvesting (e.g., dragging, changes in predator-prey relationships) also play a role in habitat loss. Declining marine water quality is a priority concern, and the three states are preparing coastal nonpoint-source pollution control strategies in response to this threat.

Among the actions that could be taken to improve the use of science in coastal policymaking, members of this issue group identified the following: (1) develop area-wide comprehensive planning programs for all sectors of the coast, (2) consider the use of a National Environmental Policy Act-like approach to integrate science into the decisionmaking process, (3) involve stakeholders in the prioritization and selection of research activities, and (4) evaluate the success of management programs in incorporating science. Specific suggestions focused on using the council as a vehicle to implement strategic planning in the region and increase public understanding of the issue.

Protecting Regionally Significant Terrestrial and Marine Habitats—Gulf of Maine habitats continue to degrade. For those habitats that support transboundary species, the council can play a vital role in stemming this degradation. For three years the council supported an analysis of regionally significant species and developed a list of 150 plant and animal species to identify priority habitats.

Among the factors that need to be addressed to improve the use of science in coastal policymaking, the issue group identified the following: (1) legal and institutional structures that tend to focus on single issues, (2) specialization among scientists that hinders information flow, (3) lack of adequate information about the location and extent of priority habitats that hinders effective management, and (4) a lack of innovative ways to craft solutions in response to complex problems. Suggestions for improving the use of science in coastal policymaking included: (1) incorporating value systems in setting priority habitats, (2) developing a habitat classification system, (3) strengthening the institutional relationships that are being fostered by the council and RARGOM, (4) coordinating and expanding data acquisition efforts on habitats, (5) improving access to habitat information, and (6) developing consistent approaches to managing Gulf of Maine coastal and marine habitats between the United States and Canada.

Using Indicators of Environmental Quality—The region's science and management community has embraced the use of indicators of marine environmental quality. Primary objectives of this regional initiative include:

- assessing the status and trends of conditions in the marine environment by monitoring appropriate indicators of change in environmental quality, especially those that will allow identification of the early stages of change, and
- assessing existing levels, trends, and sources of toxic compounds, as well

as the economic impacts of acute and chronic exposure of humans to toxic compounds transmitted through marine foods and through water contact.[9]

The issue group identified the following as necessary for improving the use of science in coastal policymaking: (1) communicate the results of monitoring programs more effectively, (2) provide appropriate planning horizons for the development and implementation of monitoring programs, (3) involve natural and social scientists throughout the process, (4) promote dialogue between scientists and policymakers, (5) sustain current binational monitoring efforts, and (6) encourage stakeholders to be proactive in both producing and using the results of monitoring programs.

THE GULF OF MEXICO SYMPOSIUM

The Gulf of Mexico is a large (600,000 mi^2) semienclosed sea with a narrow inlet from the Caribbean Sea and an even narrower outlet to the Atlantic Ocean. The United States has an extensive shoreline on the Gulf of Mexico, which is also bordered by Mexico and Cuba. Five U.S. states border the Gulf of Mexico: Texas, Louisiana, Mississippi, Alabama, and Florida.

Resources of the Gulf of Mexico are of great regional and national importance. The gulf yields over 25 percent of the commercial fisheries harvest of the United States and supports recreational fisheries valued at $2.2 billion annually. One-half of the nation's coastal wetlands are located around the Gulf of Mexico, and these wetlands have been lost at a rapid rate during the last half of the twentieth century. Productivity in the vast majority of Gulf of Mexico fisheries depends on these wetlands and numerous shallow estuaries found along the coast. In addition, the Gulf of Mexico is one of the most active areas in the world for offshore oil and gas development and production. More than 72 percent of the oil and 97 percent of the gas produced offshore in the United States comes from the region off Louisiana and Texas. Some 45 percent of U.S. export and import tonnage passes through Gulf of Mexico ports.

A large portion of the coterminous United States drains into the Gulf of Mexico via the Mississippi, Rio Grande, and other rivers. Consequently, the coastal regions of the Gulf of Mexico are greatly affected by natural and anthropogenic variations in the flow of fresh water, sediments, nutrients, and other chemical constituents from land. Reductions in freshwater inflow due to use or diversion, the effects of changing sediment supply on coastal environments, and the excess enrichment of coastal waters with river-borne nutrients are significant environmental management issues for much of the Gulf Coast.

To facilitate joint solutions to environmental protection and resource management problems around the U.S. Gulf of Mexico, the Environmental Protection

---

[9]These are the council's objectives as stated in the monitoring plan. The plan is based on two major scientist/manager meetings held in 1990 and 1991.

Agency established the Gulf of Mexico Program in 1988. This effort now includes 12 federal agencies and five state governments working in a partnership with citizens of the region. Eight issue committees have been formed to address specific problem areas: freshwater inflow, nutrient enrichment, marine debris, coastal and shoreline erosion, toxic substances and pesticides, habitat degradation, public health, and living aquatic resources. The science-policy interactions surrounding three specific issues embedded in this comprehensive list served as the focus of discussion at the Gulf of Mexico Symposium: the cumulative impacts of offshore and coastal oil and gas development, the effects of freshwater inflow changes, and water quality and shellfish production.

Summary of Gulf of Mexico Symposium Findings

Cumulative Impacts of Offshore and Coastal Oil and Gas Development—Offshore and coastal oil and gas development has proceeded over the past 50 years in the northwestern portion of the Gulf of Mexico. Much of this development took place at a time when little knowledge about environmental impacts and planning principles existed; environmental and socioeconomic impacts were of relatively little concern. As more attention was given to assessing environmental impacts, it was shown that marine pollution from oil spills has not had significant adverse impacts offshore. The environmental effects have, however, been significant in near- and onshore areas, particularly in Louisiana's coastal wetlands, which have been channelized for access and transportation. There are other activities unrelated to oil and gas development that have impacted these environments, including changes in the delivery of fresh water, sediments, nutrients, and contaminants via rivers. These multiple, often synergistic, impacts have made assessment of the cumulative effects of oil and gas development difficult. Similarly, the social, cultural, and economic consequences (both positive and negative) of oil and gas development have been substantial for coastal communities but have occurred simultaneously with other sociocultural and economic changes.

Among the means to improve the use of science in coastal policymaking, the issue group identified the following strategies: (1) improved communication between scientists, policymakers, and implementors; (2) greater cross-cultural literacy within affected groups; (3) more research focused on "real-world" problems; (4) greater incentives for scientists to participate in the policymaking process; (5) greater tolerance of unpopular findings; (6) public education; and (7) more peer review of scientific products. Specific suggestions targeted studies of the aging pipeline infrastructure, socioeconomic impacts, wetlands restoration, offshore platform removal, regions not yet developed (e.g., the eastern gulf), and improved application of risk analysis.

Effects of Freshwater Inflow Changes—Changes in the volume of freshwater inflows into coastal ecosystems and the locations and timing of flows as a result of human activities have produced extensive effects in the Gulf of Mexico region. Three areas from around the Gulf of Mexico were explored by this issue group to better understand and compare science-policy interactions: Florida Bay, the Mississippi Delta, and the Nueces Estuary (Texas). Water has been consumed for agricultural purposes or drained from the Everglades, resulting in greatly reduced flows into the large, shallow Florida Bay. For the Mississippi River Delta, the issues relate primarily to the consequences of diverting freshwater flow from the river into the surrounding estuaries to combat saltwater intrusion and wetlands loss. The Nueces Estuary drains an arid region and thus receives limited flow subject to competing demands for agriculture and municipal water supplies.

The issue group identified eight major challenges for improving the use of science in coastal policymaking: (1) obtaining a clear statement of the questions posed by policymakers and a clear statement of the answers provided by scientists; (2) determining the preexisting conditions and realistic environmental goals; (3) understanding the role and limitations of science, specifically in determining freshwater requirements for the desired ecosystem conditions or, conversely, in predicting the effects of freshwater allocation determined by other economic or political considerations; (4) dealing with uncertainties and surprises by applying adaptive management in allocating freshwater reserves; (5) accommodating the difference in time frames between managers who want immediate answers and scientists who believe that long-term studies are needed; (6) resolving conflicting scientific analysis through greater use of scientific consensus building and peer review; (7) linking environmental and economic considerations on a common basis; and (8) encouraging a water conservation ethic.

Water Quality and Shellfish Production—Because of the low tidal range and consequent poor flushing of shallow Gulf Coast estuaries coupled with warm water temperatures and numerous human population centers, Gulf of Mexico waters are particularly susceptible to contamination by human pathogens. Public health concerns have resulted in the closure of large areas of shellfish growing waters, with considerable economic impact. Although the scientific procedures used to monitor shellfish growing areas have protected human health, these procedures rely primarily on assays for fecal coliform bacteria, tests that do not directly detect the human pathogens of greatest concern, so that the protection of public health is rather indirect. On the other hand, for a variety of reasons, fecal coliform concentrations may be elevated when there is little risk of human pathogen contamination, raising concerns that shellfish harvests are being unnecessarily restricted.

The issue group identified nine different potential barriers to the use of science in coastal policymaking: (1) too little science has been applied in the development of efficient indicators of pathogens, (2) available science is poor or of uncertain value, (3) scientific information is not communicated to managers, (4) useful scientific information is ignored by policymakers, (5) available data are not in a format readily usable in the decisionmaking process, (6) other considerations outweigh the scientific information, (7) the public fails to understand the scientific facts or policy processes, (8) scientific information appears uncertain in contrast to other information, and (9) the complexity of policy and science considerations requires a diversity of participants in the process.

Specific suggestions for improving the policymaking process were developed, including involvement of all the stakeholders, education of scientists and policymakers about each other's methods and needs, translation of scientific results and their limitations so that they can be comprehended readily by policymakers and the general public, and increasing the understanding of all participants about the backgrounds and biases of the participants.

ADDRESSING CUMULATIVE IMPACTS

Some aspect of the cumulative impacts of coastal development or human uses of the coastal ocean was considered in all three symposia: coastal development in California, land and water use in the Gulf of Maine, and offshore and coastal oil and gas development in the Gulf of Mexico. Furthermore, virtually every other issue considered in the regional symposia relates to cumulative impacts. Given the inexorable pressures of population growth, technological advances, and economic development, coming to grips with cumulative impacts is perhaps the most compelling challenge confronting the science and policy communities.

No ready solutions or easy approaches for addressing the complex problem of cumulative impacts emerged from the symposia, but some common issues were identified. First, a shared understanding must be achieved among scientists and policymakers about what constitutes cumulative impacts. Second, improved methods for evaluating cumulative environmental impacts must be developed and applied. Third, the capacity of existing governance arrangements to manage such impacts effectively must be enhanced.

The following definition garnered acceptance at the symposia, focusing on aggregative effects of incremental actions:

> Cumulative impacts are those that result from the interactions of many incremental activities, each of which may have an insignificant effect when viewed alone, but which become cumulatively significant when seen in aggregate. Cumulative effects may interact in an additive or synergistic way, may occur onsite or offsite, may have short-term or long-term effects, and may appear soon after disturbance or be delayed. (Dickert and Tuttle, 1985)

Cumulative impact assessment refers to specific ways in which the process of accumulation of effects and their environmental and social consequences are identified and evaluated. Such assessment involves identification of causal connections between activities and effects and delineates the primary role of science. Management involves deciding among options for activities based on their potential contributions to cumulative impacts on environmental or societal resources. Creating effective linkages between assessment and management requires mutual understanding of management goals and environmental and social consequences by scientists and policymakers. In that regard, managing cumulative impacts poses two specific challenges for science. First, there is an implicit need to consider socioeconomic constraints and potential outcomes, including variables such as demographics, growth management scenarios, and infrastructure needs. Second, setting spatial and temporal boundaries for scientific analyses of cumulative effects is difficult. The problems inherent in setting boundaries are complicated by the need to address multiple complex issues simultaneously and requires an understanding of the various geographical units used to conduct the assessment (e.g., habitat, ecosystem, watershed, airshed, ecoregion, governmental jurisdiction). Clearly, the assessment of cumulative impacts within the context of complex and dynamic social and natural systems requires that a wide range of variables and functional relationships be taken into account.

Managing Cumulative Impacts

Although understanding and assessing cumulative impacts is challenging, their management may be even more problematic. Governmental responsibility for activities that affect the environment is fragmented, geographically and by activity or resource sector. Coupled with incremental decisionmaking, this constitutes a major institutional impediment to the management of cumulative impacts. One way to address this problem is to ensure that decisionmaking is guided by a comprehensive, long-range regional planning framework that is updated periodically and establishes specific policies for regulatory decisions to manage or avoid adverse cumulative effects. However, comprehensive plans alone will not lead to effective management. Governance reforms also are essential.

To create the impetus for needed institutional change, substantial agreement on the desired social and environmental outcomes and scenarios must be reached, or at least there must be agreement on outcomes and scenarios to be avoided. The role of both natural and social scientists is important in this regard. To be effective, an integrated comprehensive framework for managing cumulative impacts must be politically viable—it must have sufficient public support, be "equitable" in terms of who "pays" and who "benefits," and be adequately compelling to overcome resistance to change. The governance system selected must have sufficiently comprehensive and inclusive decisionmaking authority, in both geographic and temporal dimensions. It must also be endowed with adequate fiscal resources to carry out the necessarily intensive comprehensive planning, scientific research, data collection, monitoring, and public education that will be required. Finally, it must have sufficient legal authority to adapt its approaches as new information and circumstances warrant.

3

Challenges to Effective Use of Science in Making and Implementing Coastal Policy

THE ROLE AND LIMITATIONS OF SCIENCE AND POLICYMAKING

At the very heart of the issue of the use of science for policymaking is the fact that science is concerned with inquiry, description, and explanation, whereas policymaking is concerned with governance of human behavior. Science is supposed to be value-free, whereas policymaking is normative, reflecting societal values, by definition. Although it is clear that there is no value-free science, every attempt is made by responsible scientists to identify their assumptions and biases and try to minimize the latter. The policymaking process must identify value orientations and then work toward addressing community values (Hammond and Adelman, 1976).

Science should hold to the standards of objectivity, reliability, and validity. Policymaking should reflect human values, advocacy, and leadership. In this sense, scientific results can only answer policy questions of the form: What will happen to (X) if human behavior is changed in the manner (Y)? Science can never answer policy questions of the form: What *should* happen to (X)? Science can sometimes answer questions of the form: If we wish to have (X), what different values of (Y) will yield (X)?, but only after applications of the theories, methodologies, resources, time frames, and analytical capabilities available to the scientist for the particular question at hand. Social science can help us understand the distribution of beliefs, perceptions, and norms among a constituency against which various objectives, alternatives, and their impacts can be measured, but even social science cannot be normative in and of itself (Weiss, 1987).

So, for example, in the case of coastal environmental mitigation strategies in California, a scientist may predict what mitigation techniques will lead to a certain outcome but not whether or how much of that mitigation or particular outcome is appropriate. A scientist in the Gulf of Maine region may identify a reliable, cost-effective indicator of a certain condition in the environment but not whether the condition identified is acceptable. A scientist in the Gulf of Mexico region can describe the relationship between coastal development and closed shellfish waters but not how much development or shellfish closure is appropriate.

Questions that science cannot answer fall into the category of policymaking, or governance. Policymaking is the process of identifying objectives, alternatives for achieving those objectives, and their relative costs and benefits and measuring these relative costs and benefits within the context of human values. Policymaking answers questions of the form: Given that we have an objective and we know that the costs and benefits of alternative (A) will be (X) and the costs and benefits of alternative (B) will be (Y), should we do (A) or (B)? It is the governance process, with all of its requirements for planning, analysis, and public input, through which public policy decisions are made. Political processes are important considerations and are often one of the most uncontrolled and unpredictable variables in science-policy interactions.

In the case of mitigation strategies in California, if scientists communicate what strategies are available and their relative costs and benefits, the policymaking process can proceed to identify the human values against which the alternatives and their various costs and benefits may be judged. In the case of an environmental indicator in the Gulf of Maine region, if a scientist identifies a condition in the environment from a given indicator, then the policymaking process may proceed to a decision as to whether the condition indicated is desirable or undesirable, if it should be changed, and in what manner. If a scientist in the Gulf of Mexico region can describe which land and water uses result in shellfish closures, the policymaking process can then proceed to a decision concerning how much development, and how much shellfish closure, is acceptable.

The difference between science and governance is extremely important but is often ignored or confused. Scientists often feel so strongly about a particular normative position that they claim the science indicates the best way to behave. Because coastal environmental policymaking is often contentious and occurs in the midst of a complex mixture of human values and preferences, such claims are likely to confuse the discussion further, and to lead to a diminution of the credibility of the scientist (Caldwell, 1990; Jasanoff, 1990).

Science and policymaking are different from one another but complementary. The conduct of each requires different sets of expertise. The scientist must know theory, methodology, and techniques. The policymaker must know con-

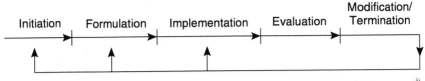

Figure 2 Stages in the policy process (Knecht, 1995).

stituencies, governance processes, and value orientations expressed as legal mandates. It is, of course, useful for each to know something of the other's trade as well, although it is unreasonable (as a general rule) to expect one to do the work of the other.

The policymaking process is composed of a number of stages (see Figure 2). In the policy initiation phase, a problem is recognized by federal, state, or local governments. In the policy formulation stage, a policy response to the problem is developed by agencies or the legislature. Policy implementation is the stage in which mechanisms planned in the policy formulation stage are made operational. In the policy evaluation stage, the results of the new mechanisms are compared with the desired outcome(s) of the policy.

Finally, policy modification/termination is the phase in which the results of the evaluation are acted on and the policy is either adapted or eliminated. Scientific input is more applicable to some stages than to others but can play an important role in each.

The policymaking process can also be viewed as a system of cultural ecology (see Figure 3), as described by Orbach (1995): "The cultural ecology of coastal environments has two broad subcomponents: (1) human constituencies of the coastal environment itself, for example, people who live on, use, or otherwise are concerned in their beliefs or behaviors with the coastal environment; and (2) humans who constitute the policy and management structures whose decisions and actions affect the behavior of the coastal constituencies defined in (1)." The cultural ecology of coastal systems is determined by the set of cultures involved in the policy process, as described in the next section.

CULTURAL DIFFERENCES

Human Culture as a Variable in the Science-Policy Interaction

All human behavior is a result of a complex interaction between culture and environment, where culture is defined as the beliefs, perceptions, and normative rules of behavior of a group of people, and environment is the total set of objects and processes with which those people interact (Harris, 1968). Culture in this sense is shared differentially among human groups—not everyone has the same

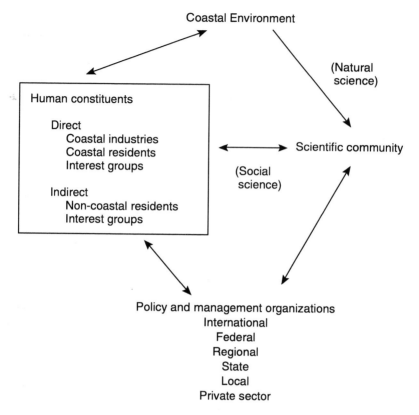

Figure 3 The cultural ecology of coastal public policymaking (Orbach, 1995).

beliefs, perceives or interprets things in the same way, or has the same normative rules of behavior. Although culture ultimately resides in the individual, certain groups share more of their culture than others, forming subcultures around linguistic, ethnic, national, professional, community, religious, and other variables. These cultural differences contribute significantly to the development of environmental policy (Caldwell, 1990).

We learn our culture, although some personality characteristics, tastes, or preferences are evidently a product of our individual genetic makeup. Most of our normative rules are taught to or internalized by us in various acculturation or socialization processes. Beliefs and perceptions are formed through a combination of the above processes in addition to our individual life experiences.

Through the acculturation process some of us become scientists, some of us become administrators, some of us become politicians, some of us become business persons, and some of us become advocates of various causes. We tend to live and work around those who have beliefs, perceptions, and norms similar to our own—hence the existence of subcultures. With respect to coastal environmental issues, all of our subcultures and behaviors interact in a complex cultural, or human, ecology that determines our societal rules of behavior, or policies (Fortman, 1990; Orbach, 1995). When we speak of the interaction between science and policy, we mean interactions among a number of subcultures, including scientists of different disciplines and employment, elected officials, legislators, administrators, business people, coastal and noncoastal residents, interest and advocacy groups, and many others (Jasanoff, 1990).

The Genesis of Cultural Differences in Coastal Policy

People acquire their professional cultures through education and training, institutional affiliation, and rewards and incentives. These lead to differences in behavior and points of view associated with the cultures of science and policy described by Boesch and Macke (1995) and shown in Table 3. Cultural differences can impede the interactions of scientists with policymakers and, consequently, the use of science in coastal policymaking. Although many individuals and groups are involved in the cultural ecology of coastal policy, we focus on two subcultures of that cultural ecology—scientists and public policymakers (defined here as legislators or administrative agency personnel).

TABLE 3 Behaviors and Points of View Typically Associated With the Cultures of Science and Policy

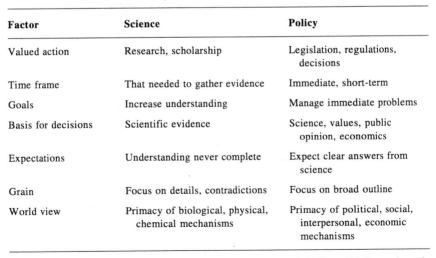

Factor	Science	Policy
Valued action	Research, scholarship	Legislation, regulations, decisions
Time frame	That needed to gather evidence	Immediate, short-term
Goals	Increase understanding	Manage immediate problems
Basis for decisions	Scientific evidence	Science, values, public opinion, economics
Expectations	Understanding never complete	Expect clear answers from science
Grain	Focus on details, contradictions	Focus on broad outline
World view	Primacy of biological, physical, chemical mechanisms	Primacy of political, social, interpersonal, economic mechanisms

SOURCE: Boesch and Macke, 1995; from *Coastal Management*, vol. 21(3), p. 189, Bernstein et al., 1993, Taylor & Francis, Inc., Washington, D.C. Reproduced with permission. All rights reserved.

Education and Training—Scientists are professionals who obtain advanced degrees, most often the Ph.D., in a specific single- or interdisciplinary training program at a college or university, thereby acquiring scientific credentials, usually in some very specific scientific domain. Scientists generally stay in school longer than the average citizen in an atmosphere that emphasizes the value of knowledge, objectivity, reliability, validity, and the scientific method. Their training institutions are somewhat insulated from society through the mechanisms designed to promote the quest for knowledge and academic freedom. University faculty instill in their students a belief in the high status of scientists and the scientific enterprise and scientists come to assume that policy must be based on science. Most problem solving in science takes the form of hypothesis testing as opposed to behavioral change.

Policymakers, although they come from a variety of backgrounds and educations, may lack scientific disciplinary focus or much education in the sciences. For example, law school, in contrast to scientific postgraduate programs, emphasizes behavioral change over hypothesis testing (Millsap, 1984). Policymakers may be people who choose to work in a world of human interaction where every new law or policy has the potential to create consensus or conflict. Rational planning, public involvement, and balanced responsiveness to constituencies and to the public trust are the hallmarks of the policymaker (Anderson, 1984).

Institutional Affiliation—There are, of course, people trained as scientists who work as policymakers. Over time, however, individuals who receive the same scientific training—and more especially others whose background and training differ—often diverge into separate subcultures based on their institutional affiliations (Fortman, 1990).

A person with scientific training who works as an administrator in a federal regulatory agency will acquire a different set of beliefs, perceptions, and norms of behavior than would a research scientist at a university because of the different requirements, contexts, and processes of their work. Individuals working in different regulatory agencies will diverge from each other for the same reasons. In the coastal area, for example, professionals at the National Marine Fisheries Service (NMFS) or the Office of Ocean and Coastal Resource Management (OCRM) of the National Oceanic and Atmospheric Administration (NOAA), the Fish and Wildlife Service (FWS), or the Environmental Protection Agency (EPA) will diverge from those in the Mineral Management Service (MMS) or the U.S. Army Corps of Engineers (Corps) because of the widely varying mandates, structures, and processes of those agencies. The mandate of the university is to investigate and educate; of NMFS, OCRM, FWS, and EPA to plan for the conservation of fishery and coastal resources; of MMS and the Corps to plan for the development of mineral and infrastructure resources.

Time Frame—For a university scientist, time frames tend to be drawn out owing to institutional factors and the infrequency of some natural events that are studied. Addressing most significant coastal issues requires long-term data and monitoring to provide information sufficient for the scientific process. Time is measured in contract and grant submission deadlines, hour-long lectures and semester-long courses, two-year article publication schedules, and decade-long research programs.

In policymaking, on the other hand, time frames and deadlines tend to be short and frequent. Regulatory development is a constant process under any given set of legislative mandates, and those mandates themselves are constantly changing. Information, power, and decisionmaking are much more hierarchical than at the university, and the policymaker will most often need to obtain data and analysis in a matter of days, weeks, or months rather than years. Thirty-day comment and response periods, controlled congressional correspondence, regulatory decisions—with the best of planning all of these are short time frame issues compared to those of the scientist.

Product Form—The products of the scientist are the results of research and the training of students. The premier product of the scientist is new knowledge, peer reviewed and disseminated to colleagues. There are, of course, many scientists who care very much about applied work—that is, science with some identifiable application to a problem or issue outside the scientific or university community— and how science is applied. Traditional academic scientific products do not, in the main, cause changes in behavior; they are not intended to.

The purpose of policymaking is behavioral change. It is our common cultural norms, as expressed through the representative democratic process and written down as laws, policies, and regulations, that constitute public policy (Nader, 1969). It is the creation of such behavioral change that is the product of the policymaker, in the form of laws, policies, regulations, and the materials, events, and processes that accompany the policy development and implementation process. An important part of the product for the policymaker is that which is communicated to the private sector constituencies and the public about the policy and policymaking process. Public involvement, for example, is an important product of the policymaking process. Public involvement is not a phrase one traditionally hears in the discussions of most scientists in their scientific work, certain social scientists excepted (Peterson, 1984). However, scientists and the public are interacting with increasing frequency, regarding the conduct of field experiments and the interpretation and application of research results to controversial environmental issues.

Cultural Conflict in Coastal Policymaking

What are the results of the existence of the different cultures and subcultures of people involved in coastal policymaking? The existence of different points of view and different interests is a major strength of the U.S. governance system, which has the structure and organization to achieve consensus among those points of view and interests. However, different cultures and subcultures also have negative effects on the use of science for policymaking. The negative effects fall into four general categories: (1) lack of understanding, (2) lack of communication, (3) lack of or misuse of each other's products, and (4) conflictual or competitive rather than cooperative interaction.

Lack of Understanding—Human ego is a powerful thing, and few things offend us and make us react in negative ways as much as the knowledge that another person does not value, respect, or understand what we are as individuals or what we do professionally. Whether it is an interaction between a fishermen and a marine biologist, an oil worker and an environmentalist, a land-use planner and a private property advocate, a social scientist and a natural scientist, or a scientist and a politician, if we interact with others with an attitude of superiority or contempt, conflict is likely. Understanding does not have to mean admiration or agreement, but simply an acceptance of the fact that the other party has a legitimate status and role in the human ecology of the policymaking process and views that must be understood in the context of that status and role.

Lack of Communication—Cultural differences, whether they stem from language, occupation, or advocacy position tend to make communication more difficult. Not only are we less likely to communicate at all with different cultures or subcultures, but communication that does occur tends to be fraught with misinterpretation or lack of understanding. The use of scientific jargon in a public presentation is one such example of this problem. A scientist and a fisherman interpreting differently the results of a trend or cycle in fish landings is another. A shellfisher and a marina owner discussing water quality is a third. Sometimes the message is not received at all; sometimes it is perceived or interpreted differently than intended (see Lampl, 1995). It is difficult, but possible and desirable, to expend the effort to open a line of communication and to be aware of the different possibilities for perception and interpretation.

Lack of, or Misuse of, Each Other's Products—It is often the case that an administrator will not know how to use the contents of a scientific report. It is often the case that a scientist will not understand the genesis or rationale for a particular public policymaking process. Private citizens will often be confused by both a scientific report and a policy process. The unfortunate response is for individuals to disengage—that is, to withdraw from the interaction or process—or simply to ignore the activity or viewpoint of others. Citizens stop attending public meetings or hearings. Scientists stop seeking funding from applied research programs. Policymakers carry out their responsibilities as best they can, assuming that the best scientific information available is that which they can interpret and use, which may be a small portion of that which scientists have produced and which may be meaningless outside the larger context. The alternative is to take the product and use it inappropriately—a scientist advocates a value position rather than simply presenting the science, a policymaker lists a report in the bibliography and uses it by reference to justify a predetermined course of action, or a citizen uses a public meeting to advance a particular constituency's advocacy agenda in the name of the public.

Conflict and Competition Instead of Cooperation—All of the above effects lead to conflict and competition in place of cooperation. They are all dimensions of the potentially negative public policy outcomes that can result from cultural differences, when those differences are not recognized, understood, and addressed.

The next section focuses on the manner in which these phenomena apply particularly to scientists in agencies, academia, industry, and nongovernmental organizations (NGOs).

SCIENTIFIC ADVISORY AND REVIEW MECHANISMS

Mechanisms for Providing Advice

Scientific information is provided to policymakers through a variety of channels, including formal reports, interactions with individual scientists, and via the public and news media. Important mechanisms also include the formal rendering of advice by scientists internal or external to the responsible agency or critical review of reports and proposals, so-called peer review. Peer review is a mechanism within the scientific community by which scientists review the work of their colleagues, usually as a step supporting a research project or publication of journal articles. This procedure serves as a check on the validity of the methods, interpretation of the data, and applicability of the conclusions drawn. While this process is not specifically directed to science-policy interactions, it provides an important quality control step in the dissemination of science and thus has an impact on public policy. For example, research on cold fusion was not subjected to peer review before the discovery was announced in a press conference and was adopted by some policymakers as a solution to the nation's energy problem. This recent example illustrates how policymaking can be affected deleteriously by the omission of peer review. Scientific advice can be obtained through at least four different mechanisms:

1. Internal Advice—The first line of scientific advice often available for designing agency programs and forming policy is from scientists who are agency employees or whose services are obtained through contracts. Internal advice may be available more quickly and tailored to answer agency questions more directly than can many forms of external advice, because internal scientists are acquainted with the agency culture and procedures. Internal advice can take the form of research findings as well as deliberative internal advisory groups. The committee did not evaluate specific means of improving the use of internal scientific advice, but most mechanisms recommended in the final chapter are applicable to both internal and external sources of advice.

2. *Advisory groups external to policymaking agencies.* External advisors can be useful to agencies and policymakers for situations in which an independent evaluation of information is needed, agencies desire to review their internal scientific mechanisms, and when it would be more cost-effective to obtain the information from outside the governmental organization. These groups may be convened by an agency from among scientists not employed by them or convened by another organization such as the National Research Council (NRC) or a professional society at the request of the agency. In the latter case, the group is typically asked to review how an agency is handling some aspect of its policymaking. There are examples at all levels of government of such external advisory functions. MMS offers a good example of the use of external committees. MMS was required by the Outer Continental Shelf (OCS) Lands Act Amendments of 1978 to establish an external Scientific Committee of its OCS Advisory Board. Members are selected from academia, technical service firms, the oil and gas industry, and government. They meet on a regular basis to help the agency set its scientific agenda and, to a limited extent, interpret the results of the MMS Environmental Studies Program. Twice, MMS has requested the NRC to review the Environmental Studies Program, which resulted in two reports (NRC, 1978, 1990c). EPA, NOAA, the National Science Foundation, and other federal agencies have one or more scientific advisory committees at different levels of the agency.

3. *Workshops.* A workshop may be convened to offer advice to an agency on a specific issue. The attendees may all be scientists, but typically the group also includes policymakers and stakeholders.

4. *Informal policy advisory groups.* The published results of scientific research performed outside an agency can provide information that is directly applicable to an agency policy decision. The information may come to the agency's attention via its own scientific professionals, outside scientists, or members of the public. With electronic mail and on-line workshops, it has become much easier to be aware of the range of information on a subject.

Why Do Scientists Participate?

If we wish to encourage cooperation between scientists and policymakers through advisory mechanisms, it helps to understand why scientists participate in such endeavors. The personality of the scientist plays a considerable role. Scientists are trained problem solvers, so they tend to be challenged by the idea that they can contribute to solving problems connected with public policy. Although not all scientists are motivated to this service, those who respond most often to such a challenge are likely to see contributions to policymaking as a stimulating extension of their professions. If they have confidence in their knowledge and its applicability to the policy questions to be addressed, they will be more willing to participate. Finally, if they believe the problems that need scientific input are significant to society, they will feel their commitment of time and effort is worthwhile.

Another reason for participation can be funding. Many research scientists fund much of their time and effort, and that of their assistants, through grants and contracts. There is a considerable lead time involved in obtaining funding, and sometimes there are gaps between funded projects. Advisory committees provide scientists with the opportunity to expand their networks and update their information on existing funding sources and fundable research. Furthermore, the possibility of funding some of a scientist's time to work on an advisory committee, recognizing that the time commitment may be great and money in relatively small units, could be a motive for serving as an adviser.

The final criterion must be that the scientist has time available and feels that she or he can afford to devote it to the purpose at hand, realizing that advisory committee work is not usually judged to be of equal value to publishing research papers in the reward structure of most scientific institutions. This situation continues to persist even though universities and agencies assert that public service is a valuable part of a scientist's career.

Impediments to Participation and Success

A number of impediments must be overcome to elicit help from scientists and to ensure that their advice can be used effectively.

Time constraints—As pointed out earlier, by the time managers realize that a policy decision must be made, there is frequently little time remaining in which to investigate the scientific bases for a decision. Under these circumstances, it is very difficult to find scientists whose schedule permits them to respond immediately. Even those within an agency may find it difficult to locate the necessary information quickly, evaluate it adequately, and respond to a request for scientific input to the decision. The case is more difficult for scientists external to an agency, if only because they must be located and recruited to the purpose before their input can be obtained.

Many of the scientific advisory bodies in government rely on volunteers, the scientists giving their time and expertise without compensation. The bigger the issues that must be addressed, the more consideration and, therefore, time that must be devoted to the matter. The more background material there is to be considered, the more time it takes to locate, obtain, and assimilate it. In most cases of volunteer members, the time required will be a major impediment to the advisory process.

Adequate staff support by the agency can make a critical difference. Committee staff can locate, copy, and distribute documents. Staff can often take the first step in preparing reports. They can set agendas and arrange meetings and meeting support. All of this can save time for the scientists involved as well as make them feel that their efforts and time are valued—that the advice received will have an effect on policy decisions.

Strong vested interests—The agency ostensibly seeking scientific advice may have so strong a vested interest in a certain policy position that it is not receptive to objective scientific advice that questions the bases of that policy position. In reality, the agency's policy position may be shaped by other legal, political, or economic considerations, but the failure to communicate these constraints honestly may lead to frustration and cynicism by the scientific advisers.

A related situation occurs when agency leaders have already formed an opinion on a subject and convene a committee to legitimize their previously held beliefs. At its extreme, this approach can skew the scope of the advisory panel's charge and its membership, result in the advisory panel being brought too late into the decision process, and can diminish the credibility of scientific information and scientists.

Lack of unanimity among scientists—Scientists frequently disagree in their interpretation of data. In fact, questioning interpretations is a necessary aspect of truth seeking in science. But if scientists on an advisory committee disagree on the interpretation of important data, it may be very difficult for agency managers to know how to use this conflicting information. There is no good solution to this problem, but its existence should be recognized by the policymakers and should not be allowed to upset the entire process. Differing biases may be the source of the disagreement (see below).

Science is not the basis for a decision—There are and will be many times when scientific information will not be the basis for a decision. Other considerations are simply more important. The desire of the public for an action may be so powerful that the policymaker is pressured to ignore scientific advice about the nature of the problem and the consequences of an action. In most cases, the process of formulating a policy requires accommodations between a variety of interests and is subject to political pressure. Thus, a decision may be made on the basis of public desire even though the action, according to scientific opinion, will not have the desired effect. In other cases, the decision will be based partly on science and partly on other considerations. It may, therefore, seem undesirable to the advisory scientists. Again, there is no way to avoid this difficulty in a democratic society, but scientists must recognize that their input is not necessarily going to be used or used appropriately.

Lack of independence of scientists—Competent scientists work in agencies, academia, business, and NGOs. They may, however, as pointed out above, represent an agency with a regulatory role. It is possible that they will find themselves in positions where their scientific judgment is affected by agency policy or where they do not have complete independence to state their opinions or accept alternative interpretations. For example, Sabatier (1995) cites a case in which scientists employed by a state fisheries agency disagreed with the interpretation of results from a scientist employed by the same state's water resources agency who had suggested a source of mortality for striped bass larvae other than that on which the management by the fisheries agency was then based. The same issue may be evident with scientists employed by businesses or NGOs with an interest in the decisions that will be affected by the scientific advice provided. Finally, it must be realized that academic scientists also have their biases. Universities are subject to the political pressures of interest groups, and if the academic scientist is funded by an agency involved in the process, his or her unconscious bias may be very similar to that of the agency scientist.

Lack of attention to the advisory committee by the agency—It can happen that an agency has a scientific advisory committee whose advice does not seem to be given sufficient attention by the agency. An example comes from the history of the previously mentioned MMS Scientific Committee. This committee criticized the scientific information that was used to support OCS leasing decisions but seemed to have little effect on the program. But when an NRC committee reported the same criticisms under different political and economic circumstances, these criticisms were used to support moratoria on leasing in Georges Bank, Florida, and California (NRC, 1989, 1991). It appears now, though, that MMS is more responsive to its Science Committee than it was in the past.

When an advisory committee proposes changes in an agency's structure or its scope of activities, the agency's reaction can be strong and negative. Such a reaction encourages many advisory committee members to withdraw and go back to their normal activities, carrying with them a general distaste for the advisory process. There are also cases, however, in which individuals are stimulated to great efforts by opposition (Scheiber, 1995).

Lack of big picture—Scientific advice will necessarily be incomplete if the purview of the advisory committee is restricted such that committee members cannot consider all aspects of the problems at hand. This may occur because of limitations in the agency's legal responsibilities or as a result of turf wars among agencies. It may occur as a result of limitations in the committee's terms of reference as given to the committee or as a result of its own failure to examine the problem completely. A complete perspective may include the entire ecosystem for the natural scientist or the whole coastal social and economic system for the social scientist.

For example, there are a variety of terrestrial and marine activities that affect spawning and nursery areas, as well as adult habitats of commercial fish species. These activities include wetland destruction, coastal pollution, and introduction of nonindigenous species. Fishery management councils have little or no authority over such activities and over habitat alterations or protection in general, reducing any attention to these matters by their statistical and scientific advisory committees. The government, not the committee, fails to define the scope of the committee's activities properly (Shelley and Dorsey, 1995).

Scientists' reluctance to extrapolate—Scientists, by training, attempt to limit their scientific conclusions to those that can be supported firmly by their data. If they extrapolate beyond this, they usually do so hesitantly and at the risk of considerable criticism from their peers. As a result, scientists frequently resist extrapolation, citing the need for further study. This typically happens at the point in the process where the policymakers need a firm recommendation based on the data available.

Scientists become advocates—Just as scientists' objectivity may be questioned if they are biased by associations or personal interests, they may also lose credibility if they go beyond *explaining*, from their discipline's viewpoint, the consequences of a certain policy to *advocating* a specific policy. Probably no scientist is without some advocacy. But when scientists become subjective advocates, their claim to strict logic, drawing from carefully bounded conclusions from properly collected data, is seriously jeopardized. There are cases in which scientists feel the evidence is so compelling that they must become advocates for reform, as in the cases of regulating DDT and antifouling paints containing tributyl tin. These cases must remain rare if scientists are to continue to be viewed as objective observers and analysts (Boesch and Macke, 1995).

Emphasis is on legal aspects and the threat of litigation—When there is large-scale environmental damage, legal actions seem to dominate and scientific advice is devalued. Actions under the Superfund law provide a long series of examples of how attention is paid to legal issues rather than to the good science needed to find the most suitable ways in which to repair damage. A good example is the *Exxon Valdez* oil spill (Wheelwright, 1994). The event offered many opportunities for scientists to learn more about damages from large oil spills and about how to clean them up most effectively. We learned what should and what should not be done. But in case after case scientists were cautioned concerning the distribution of their results, and their data were sequestered because of the legal requirements of the damage assessment process. Many scientists came through this event greatly discouraged from participating in future policy actions; others persevered.

Examples of Failures and Successes

The following examples of failures are taken from symposia discussions to show several ways in which scientific advice was not sought and/or used for coastal policymaking.

Drilling Discharge Studies—MMS and the oil industry both funded studies of the effects of drilling muds on the marine environment. These studies uniformly showed that the detrimental effects of the discharge of muds were limited to small areas immediately around the drilling rig and that there were no serious, long-term effects (NRC, 1983a). These studies were reviewed by the Scientific Committee of the OCS Advisory Board, which repeatedly recommended that no further studies were necessary or desirable from a scientific viewpoint. This opinion never had much impact. Additional studies of the effects of drilling muds were recommended by various local groups whenever new OCS leasing was proposed. The problem was probably one that should have been addressed by social scientists rather than by natural scientists, recognizing the issue as largely political and removing it from the natural scientific agenda altogether. One must conclude that the scientific analysis was misunderstood or ignored.

Northeast Groundfish—The groundfish industry in the Georges Bank/Gulf of Maine region has collapsed despite the Magnuson Fishery Conservation and Management Act, a law that established regional fishery management councils, each with an advisory scientific and statistical committee to manage the stocks. The reasons for failure are many and complex. They are related to failures to (1) take a large enough view, (2) protect fish populations and habitat, (3) use the scientific advisory committee effectively, and (4) understand the social problems involved in management of fisheries and fishermen (Acheson, 1995; Orbach, 1995).

Trinity Ledge Herring Fishery—Trinity Ledge, an area off the southwest coast of Nova Scotia, was a popular herring fishery ground. Although the need for regulation to protect the fishery was recognized, insufficient action was taken, and the area now supports only a small fishery compared to what it supported previously. The management failure was due to industry pressure against regulation and lack of consideration of the importance of the habitat for the substock of herring whose habitat was on the ledge (Chang et al., 1995). It seems clear that there was insufficient *effective* communication among industry, scientists, and regulators.

The following examples of successes and partial successes are taken from those mentioned during the symposia to illustrate that despite the difficulties advisory committees and scientific inputs have been successful in influencing policy decisions in a positive fashion.

New Hampshire Coastal Wetlands Manual—Because of the limited extent of coastal wetlands and associated habitats in New Hampshire and the intense development pressure to which they are subjected, the New Hampshire Coastal Program, the U.S. Soil Conservation Service, and the New Hampshire Audubon Society collaborated to develop a coastal wetlands evaluation manual. They involved scientists, environmentalists, citizens, and policymakers. The development of the manual is an example of a successful collaboration, although the success of its application remains to be evaluated (NRC, 1995b).

Arcata Marsh Mitigation—This mitigation program in Arcata, California, created wetlands habitat in connection with wastewater treatment. It was built on good scientific input, including pilot projects and ongoing involvement by academic scientists. This resulted in acceptance by both the public and state regulators—and an economic benefit to the local community in terms of reduced wastewater treatment costs (NRC, 1995a).

MMS Scientific Committee—This committee has provided input to MMS's Environmental Studies Program that has helped move scientific efforts in needed directions, has encouraged greater involvement of social scientists, and has involved participation by academic scientists. It is only a partial success because its recommendations have not always been heeded, as mentioned earlier.

In a paper from the California Symposium, Boesch and Macke (1995) observed that, despite the many limitations discussed above, scientific advisory committees can provide valuable services of detached criticism, public validation, and forward-looking advice. Boesch and Macke offer several suggestions to agencies and the scientists who serve on advisory committees about how to make scientific advisory committees effective (see Box 1).

INTEGRATION OF NATURAL AND SOCIAL SCIENCES

Interaction is important not only between scientists and nonscientists but also among scientists from different disciplines. In this section the committee considers issues that arise in the interaction of subcultures of scientists, particularly those from the natural and social sciences.

The Need for Scientific Integration

Environmental problems of the coastal zone have multiple attributes on physical, economic, social, and political dimensions. Because environmental problems have multiple dimensions, they cross boundaries that have traditionally

---

**Box 1
Making Scientific Committees Effective
(Boesch and Macke, 1995)**

Committee members should:

- Make an effort to focus on what is known and how this information can help the agency; they should not spend too much effort lamenting all the unfilled research needs.
- Concentrate on a future time horizon at which the committee's advice may have an effect; they should not get bogged down with today's crises.
- Avoid minutiae and details

Sponsoring agencies should:

- Have the committee report formally to the highest appropriate level within the agency (e.g., EPA's Science Advisory Board directs all of its reports to the EPA administrator); this keeps everyone honest and gives the committee a sense that its work is important.
- Avoid committees composed of individuals representing institutions or programs. Appoint individuals because of their scientific experience and knowledge, a mixture of eminent scientists (for prestige and wisdom) and younger activists (to do the work).
- Assign the committee some important but narrowly defined and doable tasks (to give it a sense of accomplishment important for sustaining interest) and at least one futuristic and relatively unbounded task (to stimulate intellectual creativity). EPA's Science Advisory Board works quite effectively with such a mix.
- Consider the committee's time priceless; use it wisely.

---

been established between the natural and social sciences. In fact, humans are integrated with natural systems in all aspects of the coastal zone, and there are few, if any, coastal environments that are not influenced by human society.

As human population has increased in coastal areas, pressures on natural resources have intensified and the number of synergistic effects among different human activities has increased. It has become increasingly clear that to manage coastal resources effectively an ecosystem approach to management is required (NRC, 1994a,b). Ecosystem management involves trade-offs between ecosystem components and requires a thorough understanding of both biological and human dimensions. Choosing what to sustain is not only a biological question but also a question of human values. Environmental problems are reflections of the conflicting goals and values of society. Humans respond to the economic, social, and political incentives that face them, but these incentives are often inconsistent with promoting sustainable use of natural resources.

The human and ecological dimensions of coastal marine resources are inextricably linked, and their linkage creates a need for the integration of natural and social sciences. The natural sciences contribute to coastal ocean policy through research that assesses the status and function of natural systems (biological, chemical, geological, and physical), the interactions of these components, and the functional relationships within and between populations and environment. The social sciences offer analogous knowledge regarding the basic attributes of economic, social, and political systems; their interactions with their environment and with each other; and the functional relationships within and between groups. An example of the need for an integrated approach is provided by the issue of freshwater inputs in the Gulf of Mexico region (NRC, 1995c). Management is faced with the challenge of balancing the need for human use of fresh water with the need for fresh water to maintain healthy estuarine systems.

The ideal role of research on coastal environmental problems is to contribute to the understanding of natural and human systems so that their interaction can be structured in socially desirable ways. Policy to adjust human behavior cannot be effective without a basic knowledge of both the natural and the human systems. If scientific understanding is incomplete, policy will fail to address coastal problems in their full dimensions.

Obstacles to Scientific Integration

Unfortunately, research in the natural and social sciences usually is not integrated in ways that will address the full dimensions of environmental problems. The obstacles to integration are many, based on differences that include history and tradition, language, world view, and incentive structures.

History and Tradition—U.S. resource management has historically proceeded on a single-resource basis. Scientific analysis of resource questions has been correspondingly specialized. The single-resource approach is the residual of an era of general resource surplus. Resources were developed singly, and the negative effects of any one resource harvesting activity on another were considered unimportant. Policies for marine resources developed on an as-needed basis in response to specific problems. Initial stages of management were centered on conservation needs. The scientific staffing of resource agencies, heavily weighted toward the biological sciences, reflects their roots in conservation concerns. Social scientists are either unrepresented or poorly represented on agency staffs. Interactions between the social and natural sciences are limited by the small number of social scientists and the infrequency of interdisciplinary research. As a result, human aspects of environmental problems are often not brought into research projects at the design stage and are more likely to be added, if at all, at the end of the research process. Academic environments are also characterized by infrequent professional interactions between natural and social scientists.

Language—All scientific disciplines develop technical language that reflects specialized and in-depth knowledge of their subject. Specialized training and the predominantly within-discipline interactions reinforce the use of discipline-specific technical language and build barriers to wider communication. The limited interactions between social and natural scientists maintain those language barriers. The result is often mutual ignorance about concepts, methodologies, and paradigms that inhibits communication and integration. An associated barrier is incompatibility of natural and social scientific data over geographic scales, time scales, and units of measurement.

World view—The professional training of natural and social scientists is based on different paradigms of ecosystems, particularly with regard to the role of humans. Natural scientists often view humans as intruders in ecosystems, whereas social scientists generally consider an ecosystem as a provider of services to humans. Natural science may focus more on the conservation or preservation of resources. Social science may focus more on the use of resources. The different world views also extend to responses to environmental variability. For natural systems the uncertainty created by variability dictates a conservative approach to use; it is better to act conservatively and underuse than to act aggressively and overuse. For human systems, however, there is an opportunity cost of foregone consumption. The human response to uncertainty is to shorten the time horizon of resource use and accelerate use in the current time period.

Incentive structures—Scientists are rewarded for specialization within their disciplines. Own-discipline publication outlets are generally more highly regarded than interdisciplinary outlets. There are numerous positive incentives that keep people within their own area of specialization, rather than interacting with scientists in a broader disciplinary area. Despite the acknowledged need for interdisciplinary research and collaboration on environmental problems, there are many factors that discourage collaboration and promote specialization. Finally, the token representation of social scientists in most resource agencies and on many advisory committees provides a further disincentive to active collaboration. A single social scientist is often expected to represent all social science disciplines with regard to a range of issues.

PREDICTION AND UNCERTAINTY

Descriptive and Predictive Science

Most natural and social science has been directed to uncovering the *causes* of change in the natural or human parts of an ecosystem, including those changes caused by human activities. Much less effort has addressed the *consequences* of these changes on the broader ecosystem, including their human components. Even less attention has been devoted to attempting to predict the effects of future activities, resource uses, or management actions on these ecosystems. Yet the issues that policymakers and implementers must address inherently require a predictive capability from science. For example, it is not good enough to know that oxygen has been depleted in a particular coastal ecosystem as a result of nutrient overenrichment; the policymaker needs to know the sources of the nutrients and how much the nutrient inputs should be reduced to achieve a certain improved condition. It is not sufficient to know that certain factors have degraded coastal habitats; the manager needs to know how these habitats can be restored. It is not good enough to know that overfishing has reduced a fish population; the manager needs to know how many fish can be harvested without additional deleterious effects on the population and whether and how a population can be maintained at optimum levels.

Most often, scientists provide information needed for the predictions underpinning policy decisions or management actions based on inference developed from retrospective analysis of the relevant changes observed. Most scientists are uncomfortable making predictions that strain the traditions of the scientific method, including the avoidance of overextrapolation of research results. The reluctance of scientists to answer policymakers' needs for unequivocal predictions and policymakers' lack of understanding of the scientific limitations for prediction are at the heart of the mutual frustration often seen at the science-policy interface.

Risk Assessment and Simulation Models

Several approaches have been pursued for predicting the outcome of policy or management decisions. Various forms of risk assessment attempt to quantify the severity or likelihood of effects or responses. In its least quantitative form, relative risk assessment has been applied to rank threats or actions in terms of severity, extent, and reversibility of effects. Relative risk assessment has been applied to rank environmental problems for the nation (EPA Science Advisory Board, 1990), a particular state (e.g., Louisiana Department of Natural Resources, 1991; Vermont Agency for Natural Resources, 1991), or community; for the long-term impacts of offshore oil and gas development (Boesch and Rabalais, 1987); and for marine pollution problems on a global scale (GESAMP, 1990). Relative risk assessment is subjective but can provide a framework for developing a consensus of expert opinion. It is not designed to make quantitative predictions of the outcome of a particular management option, but it may be useful in weighing the relative effectiveness of alternatives.

More quantitatively rigorous risk assessment is generally used to estimate the risk to human health of exposure to carcinogens or other toxicants based on a paradigm that has four phases: hazard identification, dose-response assessment, exposure assessment, and risk characterization (NRC, 1983b). A similar approach is used by EPA and other agencies to predict the effects of chemical

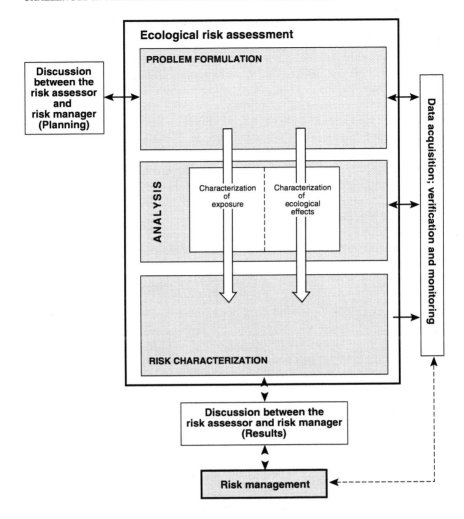

Figure 4 A framework for ecological risk assessment (EPA, 1992).

contaminants and other stressors on marine organisms or ecosystems, as depicted in the framework for ecological risk assessment shown in Figure 4. In this context, characterization of ecological effects is used in lieu of dose-response or hazard assessment, which are more relevant to chemical stressors than to the variety of nonchemical stressors that can affect components of the ecosystem. As applied to an environmental stress such as a toxic chemical, reduced dissolved oxygen, or temperature abnormalities, this involves estimating exposure concentrations by direct measurement or by modeling the dilution or fate of the stressor in the environment and experimental measurement of the effects induced at various levels of the stressor to predict the risk of prescribed effects. Such approaches frequently are employed to regulate the use of chemicals and treatment of wastes discharged into the coastal environment. Although advances are being made in ecological risk analysis of nonchemical and multiple stressors (EPA, 1992), in practice, risk assessments have several limitations, including (1) generalizations about broader effects from tests based on one or a few species; (2) scale extrapolation (e.g., from a test container to an ecosystem); and (3) relevance of the assessment to indirect and cascading effects (e.g., as manifested via food chains) (NRC, 1994a).

Another approach to prediction involves the use of a simulation model to attempt to predict outcomes based on mathematical expressions of the important functional relationships within an ecological or social system. These simulations may be rather general or highly detailed and complex. For example, population models are used extensively in fisheries management. Water quality models have evolved from the earlier hydrodynamic-sanitary engineering models to rather sophisticated ecosystem models, particularly when applied to the assessment of nutrient loading and the resulting biogeochemical responses.

For example, the Chesapeake Bay water quality model considers three-dimensional hydrodynamics, primary and secondary production, respiration, sedimentation, bioturbation, and nutrient regeneration to predict dissolved oxygen concentrations (Louis Linker, EPA Chesapeake Bay Program, personal communication). This model includes inputs from the watershed and the atmosphere and is being used to predict the effectiveness of nutrient control strategies on future oxygen conditions and living resources. Our understanding of coastal ecosystems, as well as modeling capabilities, has advanced to the point where such predictive modeling can be a very useful tool in environmental management (NRC, 1994a). The models can identify the most critical scientific uncertainties and can be useful in evaluating alternatives, if not predicting outcome precisely. In the proceedings of the Gulf of Mexico symposium, Sklar (1995) notes that "the tool for management [of freshwater inflow] will eventually be multiobjective, nonlinear mathematical models that will identify the processes that can lead to estuarine degradation and/or establish minimum maintenance levels below which biological productivity no longer supports estuarine functions such as fishery productivity, assimilation of organic and inorganic wastes, and biodiversity." It should be emphasized that any model can be no more accurate than the understanding of the relationships that went into its construction, as well as the input data used to run the model. Also, most such models are deterministic, do not express the uncertainty in predictions, and thus may provide a false sense of confidence.

Within the human sector of coastal ecosystems, predictive modeling of *economic* conditions is most advanced. However, economic models can be misleading because they often fail to quantify the environmental costs or benefits completely and do not adequately depict the delicate interactions between economic and social phenomena (NRC, 1991). New synthetic approaches developing from the emerging field of ecological economics offer some hope of dealing with the first limitation. Methods are being developed to couple economic behavior, policy options, and environmental outcomes in geographically specific models.

Decisionmaking with Uncertainty

In most cases, coastal policy is developed and management is executed without specific scientific prediction. Even in the case of the most sophisticated models (e.g., the Chesapeake Bay water quality model), considerable uncertainty in the predictions remains simply because coastal ecosystems are complex and incompletely understood and often have nonlinear responses that are difficult to model. Without an understanding of the embodied uncertainty, such models may take on lives of their own, self-defining truth (Boesch and Macke, 1995), which is at odds with the empirical "real-world" observed effects. Based on the discussions at the three regional symposia, it is clear that too little attention is paid to this uncertainty across the science-management interface—to quantifying it, understanding it, or even talking frankly about its existence. Scientific uncertainty may be used to support the positions of those advocating strict environmental protection, those advocating resource exploitation, and those seeking relaxation of environmental controls.

The "precautionary principle" was promoted by German environmentalists in the 1970s and was embraced by the North Sea Interministerial Conferences, which agreed that "in order to protect the North Sea from possibly damaging effects of the most dangerous substances, a precautionary approach is necessary which may require action to control inputs of such substances even before a causal link has been established by clear scientific evidence" (North Sea Interministerial Conference, 1990). This concept recognizes that it is sometimes a good idea to take precautionary action before scientific knowledge is complete (Cairncross, 1991). Originally applied to controls of highly toxic substances, the precautionary principle has been evoked for the control of nutrients, overfishing, and virtually every human activity affecting the marine environment (e.g., Earll, 1992). Precautionary approaches or measures are embodied in a number of recent international agreements, including the Rio Declaration on the Environment and Development and the Framework Convention on Climate Change. Although there is no agreement on the precise substantive formulation of such precautionary approaches, the concept has become central to the international community's approach to addressing environmental issues.

Without some quantification of risk, however, precautionary principles, approaches, or measures become rhetorical or, at best, difficult to apply (Gray, 1990; Gray et al., 1991). Similarly, the more familiar concept in the United States of "risk-averse decisionmaking" at least implicitly requires some evaluation of the uncertainty and severity of potential effects embodied in the concept of risk. At the same time, opposing such environmentally conservative approaches are those advocates of minimal regulation who argue that present policies are too cautious and that the resulting overregulation is unnecessarily costly. They propose the application of risk analysis and cost-benefit analysis, while asserting that an activity that could harm the environment can be continued until it is proven scientifically to be harmful.

It seems that policymakers expect greater certainty for environmental science predictions than for economic predictions. Both decisionmakers and the public are accustomed to the high uncertainty associated with economic forecasts and do not dismiss the economist for one wrong prediction. Ecological and social systems are no less complex and unpredictable than economic systems. However, the present climate does not allow environmental scientists to offer predictions without the risk of being discredited if the predictions are incorrect.

SETTING THE SCIENCE AGENDA

Who Sets the Agenda

The determination of what science is supported, commissioned, and conducted to contribute to coastal policy and management is challenging. What criteria should be used and who should make the decisions? In a time of limited public resources to support science, it is essential to plan carefully and to consider the important variables that influence the potential for success in crafting science plans that can reasonably be expected to be carried out.

One of the first challenges confronted is determining which of all the problems confronting coastal environments and communities are of the highest priority for study. As scientific methods and understanding have advanced, more questions emerge. For example, advances in analytical chemistry have allowed the measurement of contaminants at lower and lower concentrations. Coupled with this is the discovery of very subtle, nonlethal impacts of some toxic substances on marine organisms. For example, tributyl tin used as an antifouling agent in marine paints can now be assayed in seawater at the parts per trillion level, and even at these low concentrations it has been shown to affect the sexual development and reproduction of marine animals (Goldberg, 1986). As new detection methods are developed, scientists will undoubtedly continue to discover new pollution problems involving plant nutrients, environmental estrogens, algal toxins, plastics, and pathogens. But which of these threats are greater to ecosystem integrity, biodiversity, living resources, and human society? The debate is often too heavily influenced by advocacy from scientists, managers, interest groups, or by the public's concerns, which may or may not be commensurate with scientifically documentable risks.

In-depth pursuit of these problem areas is bound to be limited by finances and available personnel. Should the criteria for allocating resources for scientific activities be primarily economic? Many marine scientists argue that there may be a growing trend toward eutrophication in coastal waters through the discharge of plant nutrients in agricultural, residential, and industrial wastes. Studies in the North Sea, the northern Adriatic, and some coastal waters of the United States have documented major and large-scale changes in coastal ecosystems and related resources over periods of decades. There is the haunting possibility that changes in the nature of the base of the food chain will alter the abundances of commercially valuable fish and shellfish. Scientists point out that the number of variables that should be measured to follow the course of eutrophication is great and includes dissolved oxygen, nutrients, chlorophyll, and the nature of phytoplankton communities and their rates of production. Furthermore, they argue that understanding such complex phenomena requires long-term and rather basic studies. If economic impact is a factor, how does one balance the potentially large, but difficult to quantify, economic impacts with the substantial and rather open-ended commitments of resources likely to be required for research, monitoring, and modeling?

On the other hand, should those pollution problems that affect human health, primarily through the consumption of seafoods, be accorded a high ranking for support? On such a basis, algal toxins, pathogens, and the possible effects of environmental estrogens would merit greatest attention. Some novel monitoring procedures might be initiated—for example, the use of maricultured or genetically engineered organisms as sentinels or biomarkers for pollution or satellite mapping of the areal extent and frequencies of exotic algal blooms.

Coastal zone policy must be continually assessed to ensure that it is both beneficial and cost effective. It must be better able to put problems in perspective, on the basis of science, as knowledge advances. Governmental agencies have a tendency to avoid introspection. An illustrative example involves heavy metals in the marine environment. Heavy metal concentrations have increased over the past century in the waters, sediments, and organisms of some areas (although decreases have also been noted recently (Owens and Cornwell, 1995)). However, only three metallic compounds have been involved in serious pollution episodes (i.e., causing serious environmental or human health effects): methyl mercury in the Minimata Bay epidemic in Japan, tributyl tin in mollusc reproduction throughout the world, and copper linked to organic material in the "greening" of oysters in Taiwan. These episodes had certain unique qualities: the metals were chemically linked with organic material, and in two of the cases the events were discovered at maricultural facilities (Goldberg, 1992). Still, programs that monitor metals (e.g., NOAA's Status and Trends Program) analyze up to a dozen elements. Measurements are regularly published in agency reports and journals, but almost none of the metals have had environmental or human health effects that have been ascertained. How do we reallocate resources to more important uses?

In addition to the assessment of relative risks, the potential that new scientific knowledge could help resolve poorly understood problems, management operations, restoration, or policy development should be considered. Some policies may reflect firm social or political attitudes and may not be very susceptible to influence by new scientific information. Other policies may only be influenced by the long-term accumulation of knowledge rather than by research focused on a particular question. In one attempt to include such considerations in an assessment of science priorities, Boesch and Rabalais (1987) compared issues regarding the long-term environmental effects of offshore oil and gas development based not only on severity, duration, and reversibility of effects but also the likelihood that scientific knowledge could be improved significantly such that it would affect policy and management. Several of the highest-priority issues identified in that process were not, at that time, receiving much research support. Other recent assessments of priorities for coastal science (NRC, 1994a; National Ocean Service, 1995) have also, at least implicitly, included such considerations of how new knowledge could help resolve environmental problems.

Once priorities are established among the various problem areas, there still remains the challenge of defining the research or monitoring activities that will provide the appropriate scientific knowledge. Because of the complex interactions within coastal ecosystems and between these ecosystems and human society, this is not an easy task. Here, the roles of the research manager interfacing between the policymakers and implementers and the scientific practitioners and scientific advisory committees become very important. They need to bridge the gap between scientists, who provide innovation but may be primarily interested in advancing knowledge, and managers, who need answers quickly but may be wary of taking risks on innovative science. This gap was described as "What's the answer? What's the question?" by the issue group that addressed issues related to changes in freshwater inflows at the Gulf of Mexico symposium (NRC, 1995c).

Factors That Should Be Considered in Setting the Science Agenda

The pressures of population growth, the needs of economic development, demographic change, competing and conflicting uses, judicial decisions, politics, competition for scarce public fiscal resources, the demands of a diverse and fragmented society, and the increasing complexity of the issues raised (e.g., addressing cumulative impacts) will inevitably drive agenda formulation and implementation. Many of these factors are discussed in greater detail elsewhere in this report and involve the following dynamics, principles, and assumptions:

- the importance of timely and effective interaction between science and policy in all phases of policy formulation and implementation;
- the differing needs and dynamics of the science and policy cultures (see pp. 29-33);

- the importance of ingenuity, innovation, and peer review;
- the relative value of fundamental and applied research and of reactive (e.g., damage assessment) versus proactive (e.g., predictive modeling) scientific activities;
- the meaningful and appropriate involvement of stakeholders in the development and support of the science agenda;
- the compelling need to achieve programmatic and logistical efficiencies and effectiveness; and
- emerging approaches such as "integrated" and "adaptive" management (see pp. 59-62) and strategic thinking relative to "place-based" policymaking (e.g., ecosystem and watershed planning).

With these factors in mind, a science agenda can be developed and implemented. Obviously, the principles involved in "setting" the agenda will differ from those that determine the degree to which the science that emerges influences policymaking. The scientific and policy communities and, in appropriate cases, the public must be involved in setting the agenda. In turn, each community will be influenced by its respective constituencies or motivational forces, such as the expectations of academic institutions; pressure from those who will benefit economically from the outcome; personal interests, goals, and objectives; the expressed desires of influential interest groups; and the perceived need to address contemporary environmental and societal problems.

Role of Fundamental Research

Although the focus here is on setting the agenda for policy-relevant science, it should be recognized that advances in policy-relevant knowledge also depend on advances in understanding derived from fundamental or basic research (NRC, 1992b). By definition, such research is not focused on solving an immediate practical problem and thus it is difficult to predict if and how the research results might eventually be useful. Nonetheless, our understanding of the effects of human activities on coastal ecosystems and societies has advanced considerably as a result of fundamental research, from advances in measurement capabilities, studies of basic biology and geochemistry, and theoretical studies. Fundamental research efforts organized to pursue a specific theme, such as the Land Margin Ecosystem Research and Coastal Ocean Processes programs funded primarily by the National Science Foundation, are now making major contributions to our understanding of estuaries and continental shelves.

More effort is needed in the interpretation of fundamental science results for use in policymaking. Perhaps the most effective means of such integration is by coastal scientists who are engaged in both fundamental research and policy-relevant scientific activities, although such individuals are a rarity. They are able to extend the results of more applied, and often more descriptive, research by bringing in the understanding of processes resulting from fundamental research. Furthermore, the availability of large amounts of descriptive information from monitoring studies provides a context for the formulation of hypotheses and the interpretation of fundamental research results. For example, neither monitoring measurements nor research experiments alone was sufficient to answer the question posed by managers in the Chesapeake Bay: "If water column nutrient inputs were reduced by 40 percent, how long would it take for the nutrient levels in the bay to respond, considering that there are large amounts of nutrients stored in bottom sediments that would continue to leach out?" But with the plentiful background information provided by 10 years of monitoring, appropriately designed research experiments were able to demonstrate that this "sediment memory" effect would last only about two years (Boynton et al., 1995).

National and Regional Needs and Roles

Although national policies may set the general management framework or establish certain standards, the policies that most affect coastal ecosystems, resources, and societies are implemented at the regional, state, and local levels. For example, coastal construction, land development in the watershed, agricultural practices, harvesting of most resources in territorial waters, and discharge permits are managed primarily from state capitals, county seats, and city halls rather than from Washington, D.C. Furthermore, the increased emphasis on integrated, place-based management raises additional responsibilities for state and local governments and multijurisdictional regional programs. Yet it is the federal government that bears the primary burden for supporting coastal science. How can it be assured that this science is relevant to scales ranging from regional to local while at the same time avoiding unnecessary duplication of these efforts, which the nation cannot afford?

Some national scientific efforts are undertaken to guide national strategies for environmental protection or coastal management. For example, NOAA's National Status and Trends monitoring program includes chemical and biological measurements made with standard techniques at a relatively sparse array of sites around the country. This program has identified regions of the country that have high concentrations of certain chemical contaminants or a high incidence of maladies of marine organisms and has demonstrated certain trends. However, these results are not used much in environmental management at the regional scale, in large part because the sampling density is too sparse to assist in management on these smaller scales. EPA has also undertaken an estuarine component of its Environmental Monitoring and Assessment Program (EMAP) in the Mid-Atlantic and Gulf of Mexico regions. Again, because this program was not designed with more local-scale management in mind and frequently is not coordinated with existing local or regional monitoring programs, EMAP results have not been used much by management programs that focus on a particular estuary or state. An earlier NRC report (NRC, 1990b) recommended integration of these national monitoring programs and the inclusion of existing or new regional monitoring programs of greater intensity within the national network as a way to meet the needs for environmental management on both national and local scales, but this has not been accomplished. Similarly, NOAA's strategic assessments of coastal data around the nation have produced reports that are very useful in revealing national patterns and trends but that are not seen as particularly useful by state and local coastal managers, who require more detailed information. An exception is in relatively unstudied areas, such as the Barataria-Terrebonne estuary (Rabalais et al., 1995), where such data may constitute the only information related to chemical contaminants.

This problem of monitoring at appropriate scales presents a difficult challenge. To meet this challenge will require federal involvement in selected regional scientific programs and improved synthetic understanding by both scientists and managers so that knowledge can be better extended from one region to another. Another improvement needed is better availability of state and federal data.

DEALING WITH COMPLEXITIES IN THE COASTAL DECISIONMAKING PROCESS

The traditional paradigm for managing coastal and ocean resources is sector-by-sector management of specific resources like fisheries, oil, and gas through relatively well-delineated authority by state or federal governments and involving a limited number of participants, primarily those most directly affected. An important exception to this approach is the Coastal Zone Management Act, which integrates management of resources to some extent.

There has been a growing realization nationally and internationally, particularly in the past decade, that such an approach is no longer applicable in many cases. Many of the issues facing coastal areas are transboundary in nature and involve multiple jurisdictions and multiple participants with diverse interests and perspectives. Examples include management of estuaries bordered by several states and management of nonpoint sources of pollution. In such examples we have seen the involvement of a wide array of participants, some of them relative newcomers to decisions about natural resources—state, federal, and local agencies; nongovernmental organizations whose numbers have grown in size and complexity in recent years; user and industry representatives not only from the ranks of the most affected use/industry but also from other related uses and industries; scientists (primarily from the natural sciences, but increasingly from the social sciences as well); and members of the public. These interactions are often adversarial, and typically there are not well-established fora or mechanisms for conflict resolution.

Such examples have drawn attention to the need to consider the effects of activities of one sector (such as agriculture) on other sectors (such as fisheries) and on the environment (fisheries habitat), to find new ways of resolving conflicts in multiple-actor and multiple-jurisdiction situations, and to adopt management approaches that are adaptive—that anticipate problems and issues and incorporate "learning" about the natural and socioeconomic environments and the performance of government programs into the management process (see pp. 61-62). As eloquently stated at the 1992 United Nations Conference on Environment and Development, achieving sustainable development of oceans and coasts will require new management approaches, that are "integrated in content and anticipatory in ambit" (UNCED, 1992).

To devise more integrated and adaptive approaches to management that incorporate a strong interface between science and policy, we must first understand the complexity of perspectives that are typically present in multiple-jurisdiction, multiple-actor situations.

Policymakers and Policy Implementers at Different Levels of Government

Policymakers, including Congress, state legislatures, regional bodies, county boards, and city councils, are responsible for responding to environmental problems by designing policies and programs, generally in the form of legislation. Policy implementers include federal agency officials, state agency officials, and regional, county, and city officials. Implementers are responsible for putting legislation into practice by developing regulations and monitoring and enforcement programs, also with public input.

Scientists

Scientists are employed in academia, government, industry, and nongovernmental organizations. Scientists may play different roles—as purveyors of objective information, authority figures, advocates and antagonists, and/or cooperators (Boesch and Macke, 1995; Sabatier, 1995). Policymakers and the public can become confused when scientists oppose each other because of differing interpretations of their own and others' research results. This situation leaves policymakers in a quandary, may paralyze decisionmaking, and may considerably diminish the role that science plays in the policy process. Solid data and analysis may be dismissed because of criticism of individuals unqualified in the scientific field in question (NRC, 1995c). The academic peer review system has as its major goal the maintenance of "good" science, but it is not always apparent to individuals not familiar with the issue (including other scientists) how to compare the quality of science and statements associated with opposing scientists.

Users/Industry

Expansion in the scope of coastal issues has meant an expansion in the number of users affected by and involved in the policy process. Users have become increasingly organized and active. Many coastal industries and user groups support regional and/or national coordinating entities such as associations or institutes. Examples include the American Petroleum Institute for oil and gas issues, the National Fisheries Institute for commercial fishing issues, and the American Sportsfishing Association for recreational fishing issues.

Nongovernmental Organizations (NGOs)

The number of NGOs active in coastal decisionmaking has grown significantly in recent years. NGOs play an important role in bringing new issues to light, educating the public, contributing to the policy development process, and acting to monitor the process. NGOs may pursue different interests (e.g., environmental, business) and vary in the extent of their interactions with the public, scientists, and policymakers. NGOs increasingly enlist scientists in their work, and their impact has increased. Examples of national and international NGOs that focus on coastal issues include the American Oceans Campaign, the Center for Marine Conservation, the Environmental Defense Fund, Greenpeace, and the National Coalition for Marine Conservation.

The Public

Members of the public participate in the decisionmaking process either as members of organized interests (users, NGOs) or as individuals taking advantage of the many opportunities for public participation provided by U.S. environmental laws. It is public values and perceptions, in an aggregate sense, that provide policymakers with direction and goals. The public has opportunities to influence policymaking through contact with legislators and by providing input during comment periods associated with new legislation.

There is a tendency for scientists and managers to believe that complete knowledge and understanding on the part of the public will be followed by agreement with the scientific and management decisions. Therefore, scientists and managers may believe that if a community is not happy with a management regime or decision it is because community members do not understand the issues. However, agreement and compliance do not necessarily follow understanding. The incorrect assumption is that an educated public is an agreeable public.[10]

---

[10]Robert Bowen, University of Massachusetts. Remarks given at the Gulf of Maine Symposium on Improving Interactions Between Coastal Science and Policy, Kennebunkport, Maine, November 2-4, 1994.

It is not enough simply to inform the public about all the information used in the policy process. The public must have the opportunity to analyze the information and to voice its concerns and desires.

In recent years, public understanding of science has been increasing and there are many instances where citizens, individually and through organized efforts (such as citizen advisory committees), have played important roles in defining and overseeing the conduct of scientific studies aimed at resolving problematic coastal issues. For example, each National Estuary Program includes citizen advisory groups, and those groups, in addition to more general public input, are integral to the development of comprehensive conservation and management plans (e.g., see Albermarle-Pamlico Estuarine Study, 1995).

The News Media

One of the most important conduits to policymakers and implementors is the popular and semipopular print and electronic news media. These media provide information directly, help shape public opinion, and affect policymakers' impressions of public opinion. For example, both in the case of ocean dumping in the New York Bight and offshore oil and gas development off California, Florida, and New England, the news media helped develop public fear that exceeded scientific assessment of the risks (Freudenberg and Gramling, 1994), leading to congressional bans or moratoria. If certain aspects of the issues are reported out of context or without full media understanding, those reports will play on the public's fears and emotions. Sensationalistic reporting tends to create much public sympathy over emotional issues, such as wildlife management, and may lead to clouded perspectives and calls for unreasonable, inefficient action.

On the other hand, the media can also be very effective in educating the public and policymakers about rather complex environmental issues and marshalling support for action against more insidious threats. A good example concerns nutrient overenrichment and oxygen depletion in the Chesapeake Bay. The media need to be targeted as an important participant in coastal management that needs better access to scientific information, so that it will not sensationalize environmental issues. The scientific community has a responsibility to communicate with the media and to encourage the reporting of issues in the proper context and with the correct amount of neutrality. With understandable scientific information, the media will have a basis on which to build a responsible role as an information provider to the public and to policymakers and implementers.

One instance where media access to scientific data has been important is in the management issues surrounding Boston Harbor (Connor, 1995). The Massachusetts Water Resources Authority published a State of the Harbor report that put issues and scientific information into lay terms, and this has led to increased media coverage of the issues. Similarly, the tabloid-style *Bay Journal* (see Figure 5) presents scientific information about the Chesapeake Bay and its watershed to the public and news media in an approachable and understandable form.

BAY JOURNAL

ALLIANCE FOR THE CHESAPEAKE BAY

| Vol. 4 No. 10 | A public education service of the Chesapeake Bay Program | January-February 1995 |

Virginia ponders withdrawal from coastal fish panel

By Karl Blankenship

BARELY a year after Congress enacted a law requiring Atlantic Coast states to comply with jointly developed fish management plans, Virginia officials are considering a rebellion.

Under a bill soon to be considered by the state General Assembly, Virginia would quit the Atlantic States Marine Fisheries Commission, which sets catch limits for fish that migrate across state borders.

The bill, sponsored by Del. W. Tayloe Murphy Jr., D-Westmoreland County, was favorably reported out of committee last fall.

Its supporters object to the commission's ability to impose fishing restrictions on individual states. Until Congress passed a law in 1993 providing federal enforcement power for the commission's plans, states could ignore ASMFC catch limits. Proponents of the federal law argued that by ignoring the plans, states were allowing coastal fish species to be overharvested.

The issue become more heated last year when the ASMFC refused a request from Maryland and Virginia to increase the amount of striped bass they could harvest, a move that angered some commercial fishermen.

Some have argued that the Bay states, where almost all striped bass are spawned, were being unfairly outvoted by Northern states who were bound by even more restrictive catch limits.

Virginia Gov. George Allen raised the issue at the Oct. 14 meeting of the Bay Program's Executive Council, saying, "We feel that we have competent expertise to manage our fisheries wisely. We do not need intrusive federal mandates that are based on politics rather than science.

"We're going to fight for our rights — if we feel we are right with our science and our evidence — when we think that others are trying to prescribe unbalanced fishery management plans that are inappropriate for our particular needs and circumstances," Allen said.

A spokeswoman for Virginia Natural Resources Secretary Becky Norton Dunlop said the Allen administration does not have a position on Murphy's specific legislation, but said that in principle it is "unhappy" about federal mandates on the state.

If Virginia decides to withdraw from the commission, it still would have to follow restrictions set by the commission.

Please see ASMFC — page 4

Biologists use a seine net to collect a fish sample from the Mattawoman Creek.

Rating Chesapeake rivers

Scientists seek 'index' that will let fish to speak for ecosystem

By Karl Blankenship

AFTER hauling a 100-foot seine net around a semi-circle from the beach, two biologists dragged the net out of the Mattawoman Creek and onto dry land.

Inside the net, more than a thousand fish — ranging from fingernail-size to several inches in length — flailed about.

A half dozen biologists swarmed around and began dividing up the catch.

Lines of bluegills, piles of striped bass, and groups of bay anchovy began to form on the sandy beach. "Anyone doing white perch?" one sorter called out, holding a few specimens in her hand.

They counted the total number present for each species, measured the largest and smallest among them, and divided them by age or "year class." They shouted the results to a record keeper: "White perch class, 55." "Maximum white perch, zero year class, 80."

The counted fish were tossed into a cooler of river water so they could later be returned to the Mattawoman.

This would go on for nearly half an hour. Then it would be done again. And then, the whole process would be repeated at four more sites in Mattawoman, a tidal fresh water river that enters the Potomac River about 20 miles south of Washington, D.C.

Offshore at each site, a boat hauled a trawl net along the bottom for five minutes. Then, biologists would inventory whatever turned up — always far

Please see INDEX — page 8

Figure 5 Cover page of January-February 1995 issue of the *Bay Journal*.

INTEGRATED AND ADAPTIVE MANAGEMENT

As discussed in preceding sections, coastal environmental and resource policies and management approaches have frequently focused on specific activities, resources, or environmental media and thus have failed to adequately reflect the linkages among them. Moreover, environmental and resource policies have, in part because of the failure to take into account this complexity, often not achieved the desired objectives or have had unanticipated outcomes. To address these twin problems of complexity and unpredictability, two important management concepts have emerged: integrated management and adaptive management.

Integrated management attempts to encompass the complex scope of multiple sectors, jurisdictions, and actors to achieve management that cuts across users, agencies, geography, resources, and disciplines. Adaptive management, aimed at the temporal dynamic aspect of management, is an approach that incorporates, on a continuous basis, learning about natural and social environments and about the performance of government programs in the management process.

Meaning and Approaches to Integrated Management

There has been considerable work in recent years in defining the major characteristics of integrated management in the context of coastal areas; see, for example, Sorensen and McCreary (1990), OECD (1991), Bower (1992), Chua (1993), NRC (1993), and Van der Wiede (1993). Although different authors emphasize somewhat different aspects of integrated coastal management (partly as a result of diverse disciplinary backgrounds and partly as a reflection of the authors' varying experiences acquired in work on integrated coastal management in different parts of the world), there appears to be growing consensus on the outlines of a general model of integrated coastal management. This is evident in recent work by the World Bank, the United Nations Food and Agriculture Organization, and the United Nations Environment Programme in the preparation of international guidelines for integrated coastal management.

There appears to be clear consensus that integrated coastal management represents a continuous and dynamic decisionmaking process. Integrated coastal management is a process by which decisions are made regarding the use, development, and protection of coastal areas and resources. It recognizes the distinctive character of the coastal zone—itself a valuable resource—for current and future generations. The goals of integrated coastal management are to attain sustainable development of coastal areas, to reduce vulnerability of coastal areas to natural hazards, and to maintain essential ecological processes, life support systems, and biological diversity in coastal areas. Integrated coastal management has multiple purposes in that it analyzes implications of development, conflicting uses, and interrelationships between physical processes and human activities and promotes linkages and harmonization between coastal activities among different sectors.

Authors differ in terms of what areas, resources, and activities they include under the aegis of integrated coastal management. In terms of areas, integrated coastal management generally must include both coastal lands and coastal waters because of the important reciprocal effects of processes and activities in these two areas.

Compared to sectoral entities and processes that tend to be concerned only with one use or resource of the coastal environment, a well-functioning integrated coastal management process is expected to perform three important roles: (1) as an area-based (rather than a single-use or single-resource-based) process, integrated coastal management has a special role in planning for the uses of a coastal area in the present and into the future, in harmonizing and balancing existing and potential uses, and in providing a long-term vision; (2) in promoting particular appropriate uses of the coastal zone that may need some special encouragement (e.g., marine aquaculture); and (3) stewardship of the ecological base of coastal areas and the promotion of public safety in areas typically prone to significant natural, as well as man-made, hazards.

Achieving integrated management in the coastal context is especially complex because several major dimensions of integration need to be addressed: (1) integration among sectors (among coastal sectors, for example, fisheries, and tourism) and between coastal sectors and other land-based sectors such as agriculture (intersectoral integration); (2) integration between the land and water sides of the coastal zone (spatial integration); (3) integration among levels of government (local, state, regional, and national) (intergovernmental integration) and among agencies within each level of government (interagency integration); and (4) integration among disciplines (natural sciences, social sciences, and engineering) and policymaking/implementation (science/policy integration).

Efforts to achieve policy integration are often most successful when incentives are utilized to entice government agencies to cooperate. Becoming involved in interagency relationships implies that an agency may lose some of its freedom to act independently and must devote scarce staff and financial resources for cooperative activities. Purposeful interagency cooperation, it would seem, will tend to take place when positive incentives to begin and maintain the interagency relationships are present. Various factors that can work as incentives for interagency cooperation are analyzed by Weiss (1987): (1) perception of a common problem that cuts across various agencies, (2) monetary incentives, (3) legal mandates, (4) sharing of norms and values among agencies on the need for integration, (5) the possibility of gaining political advantage, and (6) the possibility of reducing uncertainty.

Meaning and Approaches to Adaptive Management

Adaptive management may be defined as management systems that have the capacity to learn from their surroundings by incorporating timely information from appropriately designed sensing systems and, thus, being able to adapt to changing circumstances (see, generally, Lee, 1993). Adaptive management approaches are suggested when a capacity to cope with uncertainty and complexity is required, as is often the case in the management of natural resource systems. The conventional approach to planning and management requires a level of information "up front" that is not generally available in these cases.

Adaptive management involves the concept of learning by doing. The conduct of governance programs should be seen as opportunities to test and improve the scientific basis for action. As a strategy of implementation, adaptive management provides a framework within which management measures can be evaluated systematically as they are carried out.

A governance system that is fully "adaptive" would, in the committee's view, be one that is continuously learning from its ongoing management activities and systematically applying that learning in such a way as to make the best possible decisions. One of the keys to success, of course, will be to conduct the requisite learning ("sensing") in the right areas so as to anticipate emerging management needs. This learning must extend beyond the physical environment targeted for management attention (e.g., erosion rates or rates of sea-level rise) to include changes in the institutional, political, social, and economic environment that could affect the behavior of the governance system.

Implications for Science-Policy Interactions

Both natural and social sciences must participate significantly in efforts to achieve integrated management—the former in understanding the nature and dynamics of the natural ecosystems in question, and the latter in understanding the socioeconomic factors involved as well as the full array of players, issues, and perspectives that must be reconciled and the range of incentives and tools that can be utilized to achieve such integration. Adaptive management generally requires that scientists participate in the management process on a more intimate and frequent basis than is comfortable and in roles that are nontraditional.

Natural and social sciences are centrally involved in adaptive management:

• in the collection and analysis of systematic data regarding natural and social systems and changes in these systems and on the performance and outcomes of governance programs, and
• in developing recommendations for adaptations (changes) in management programs on the basis of the above analyses.

The information presented in this chapter sets the context for understanding the present use of science in policymaking. The factors described can either hinder or encourage effective use of science. From this background can be drawn findings and recommendations for developing improved means of using science for coastal policymaking presented in the next chapter. By identifying constraints, the committee determined that specific actions could be taken to overcome them.

4

Findings and Recommendations

Several themes ran through the symposia and the discussions of the Committee on Science and Policy for the Coastal Ocean: (1) coastal scientists and policymakers do not interact sufficiently to ensure that decisions and policies related to coastal areas are adequately based on science, (2) coastal policies tend to lack sufficient flexibility and are most often designed to manage single issues, and (3) the allocation of available resources to the application of coastal science for policymaking is suboptimal. To address these concerns, the committee developed three general recommendations:

1. Improve the interaction between natural and social scientists and coastal policymakers/implementors at all levels of government.
2. Employ integrated and adaptive management approaches in coastal policymaking and implementation.
3. Improve allocation and coordination of resources to achieve effective interaction between coastal scientists and policymakers.

The findings that support these recommendations are described separately below, as are specific suggestions under each of the general recommendations. Most of the recommendations could be applied at federal, state, and local levels. Furthermore, these recommendations are directed not only at governmental agencies and elected officials but also scientists and academic institutions, industry, nongovernmental organizations, the news media, and the public. The recommendations and who should carry them out are summarized in Table 4.

We recommend that the National Oceanic and Atmospheric Administration,

TABLE 4 Summary of Recommendations

Recommendations	Federal Agencies	Congress	State Agencies
1. Improve Interactions Between Scientists and Policymakers A. IMPROVE MECHANISMS			
Create mechanisms for external scientific review of programs	X	X	X
Involve stakeholders in planning and application of policy-relevant scientific research	X		X
Form multidisciplinary regional task forces to address complex issues	X		X
Encourage groups of scientists to develop plans for strategic research			
Reevaluate legal requirements that may hinder communication exchange		X	
B. ENHANCE COMMUNICATIONS			
Policymakers and implementers should specify their information needs	X		X
Summarize results of scientific research in lay language and disseminate widely, including through electronic information networks	X		X
Assist the media in understanding and disseminating scientific findings	X		X
C. BUILD CAPACITY			
Assess recent experiences with science-policy interaction as possible future models (e.g., NEPs)	X		X
Agency scientists should be encouraged to maintain their expertise and stay knowledgeable about current developments in their fields	X		X
Enhance science-policy training by: • enhancing natural marine science programs with social science and policy studies • enhancing marine-oriented social science programs with natural science training • creating or enhancing programs that train "science translators"			
Create consortia for strategic research	X		X
Provide academic rewards to encourage scientific involvement in policy development and implementation			
Encourage scientific involvement at all stages in the development of coastal policies	X		

State Legislatures	Local Authorities	Scientists	Universities	The Media	NGOs	The Public
X	X					
	X					
	X					
		X	X			
X						
	X					X
	X		X	X	X	X
	X	X			X	X
	X					
	X	X				
			X			
X	X	X	X		X	X
X			X			
X				X		

TABLE 4 Continued

Recommendations	Federal Agencies	Congress	State Agencies
2. Employ Integrated and Adaptive Management in Coastal Policymaking and Implementation			
Develop and employ integrated and adaptive management approaches to policy development and implementation	X		X
Allocate resources to implement existing plans to achieve integrated and adaptive management	X	X	X
Evaluate performance of state coastal programs through application of science	X		
Assess "state of the coast" in regular periodic reports	X		X
Improve scientific prediction, modeling, risk assessment, and measures of uncertainty			
3. Allocate, Mobilize, and Coordinate Resources			
Devote a portion of scientific research budgets to translate and disseminate scientific research	X	X	X
Promote interdisciplinary policy research teams in requests for proposals	X		X
Integrate science and policy capabilities through data sharing, colocation of facilities, and cooperative projects	X		X
Facilitate personnel exchange and staff sharing among universities, NGOs, industry, and agencies	X		X

NGOs—nongovernmental organizations
NEPs—National Estuary Programs

the Environmental Protection Agency, the Department of Interior, the Department of Energy, the U.S. Army Corps of Engineers, and other relevant federal agencies review the recommendations in this report for application at the federal level. Agencies could benefit from the recommendations through revisions to existing agency policies, programs, and practices and in the creation of new ones.

Congress should consider the recommendations contained herein in the development of legislation affecting coastal environments and their resources, particularly in the next reauthorization of the Coastal Zone Management Act (CZMA).

These recommendations could provide useful guidance to state agencies and legislatures. Authorities in states and regions could benefit from an analysis of

State Legislatures	Local Authorities	Scientists	Universities	The Media	NGOs	The Public
	X					
X	X					
X						
			X			
X	X					
	X					
	X					
X	X	X			X	X

region-specific suggestions summarized in Chapter 2 and discussed in more detail in the proceedings of the regional symposia.

ISSUE 1
INTERACTIONS BETWEEN COASTAL
SCIENTISTS AND POLICYMAKERS

Finding: Coastal scientists and policymakers do not interact sufficiently to ensure that decisions and policies related to coastal areas are adequately based on science.

Interactions between coastal scientists and policymakers have not been adequate to support the decisions and policies made for coastal areas. In many cases, federal, state, and local entities with responsibility for designing, implementing, and enforcing decisions and policies related to coastal management do not elicit independent advice from natural and social scientists and must therefore rely on their often limited internal resources and expertise. Often, scientists who are employed by government agencies are unable to maintain their expertise and find it difficult to provide all the scientific services needed within an agency.

There are few mechanisms to plan and carry out "strategic research" to support science-based policymaking and to encourage agency and external scientists to participate in policy development, implementation, and evaluation, although many federal and state agencies have established scientific advisory and review mechanisms (see pp. 35-42). Unfortunately, the mismatch in cultures and time scales between scientists and policymakers (see pp. 29-35) sometimes diminishes the effectiveness of advisory committees.

The human and other dimensions of coastal environments and their resources are inextricably linked, and their linkage creates a need for the integration of social and natural sciences. Natural and social scientists seldom interact professionally and have different traditions, languages, world views, and incentive systems (see pp. 42-45). In part, such barriers exist because there is inadequate cross-training between the social and natural sciences in graduate programs, although public policy programs increasingly stress interdisciplinary skills.

The public can have a major influence on coastal policy. Whether public influence helps solve environmental problems or hinders solutions depends on the public's level of knowledge about an issue. The public can exert tremendous political influence regardless of its knowledge of the details or scientific background of an issue. This means that the transfer of scientific knowledge to the public is at least as important as its transfer to policymakers. Means of communicating scientific information to promote understanding about coastal environmental issues, through the media and other fora, are deficient in that regard (see pp. 57-58).

Laws, regulations, and administrative and legal decisions are designed ideally to protect the environment and the public's health, safety, and rights and to manage resources wisely. Many coastal problems are regional in nature, crossing jurisdictional boundaries and having both environmental and social aspects. Such problems may require teams of experts from different sciences and different levels of government to work together. However, interactions among federal officials, state officials, and external scientists can sometimes be seriously inhibited by well-intended laws designed to ensure public access to policymaking, such as the Federal Advisory Committee Act (NRC, 1995c).

Some coastal management programs have gained experience in applying science to design and implement coastal programs. For example, the National Estuary Program (NEP) of the Environmental Protection Agency (EPA) forms broadly constituted groups of scientists, citizens, and policymakers to design programs for protecting and improving coastal environmental quality. The Chesapeake Bay and Great Lakes programs bring together similar constituencies to focus on coastal environmental issues in their regions.

Recommendation: Improve the interaction between scientists (natural and social) and coastal policymakers/implementers at all levels.

Improve Mechanisms for Focusing Scientific Attention on Coastal Environmental Issues

• Federal, state, and local entities are encouraged to create or enhance mechanisms for internal and external scientific review and assessment of their coastal programs (NRC, 1995b), including the science conducted by internal staff and contractors.

Review and advice may be solicited from standing scientific committees, peer review panels, or through other mechanisms. As agencies seek review and advice, they should keep in mind the impediments to participation of scientists in the advisory process, as well as impediments to the success of advisory activities noted on pp. 37-41. Effective use of scientific review and advice not only improves the use of science in coastal policymaking and management but also engages scientists as more active participants in coastal management programs.

• Federal, state, and local entities are encouraged to involve stakeholders in policy development, implementation, evaluation, and modification, including the planning and application of policy-relevant scientific research.

Particularly important is stakeholder involvement in the initial planning, definition of tasks to be accomplished, and identification of entities that should be involved in the process. Actions should be initiated to educate stakeholders about the availability of scientific information and the importance of using it in the coastal management process (NRC, 1995b). Experience gained from the NEPs, the Chesapeake Bay Program, and the Great Lakes Program could provide a model for stakeholder involvement in coastal management. Including stakeholders in the process will give them a vested interest in the outcome (NRC, 1995c) and will reduce uncertainty about the range of outcomes desired by the public (NRC, 1995b). Human motivation and responses should be included as part of the social systems to be studied and managed (McKinney, 1995).

• Federal, state, and local entities should encourage the formation of regional problem-solving task forces or groups to address coastal problems that cross subject areas, legal jurisdictions, and policy sectors, using, when relevant, an ecosystem approach.

Participants in the Gulf of Mexico symposium suggested that, to deal with the interactive issues of oyster production, water quality, and human health, health officials and environmental quality officials should work together to integrate their efforts (NRC, 1995c). Participants in the California symposium suggested the formation of a blue-ribbon panel, including scientists, policymakers, and resource agency personnel, to "define the information needed for decisionmaking over the long term, define a research agenda to obtain this information, and assess and synthesize ongoing science" in the area of habitat mitigation (NRC, 1995a). Sabatier (1995) suggested the formation of specialized fora to promote interactions, which should be sufficiently prestigious to induce professionals from different advocacy coalitions to participate and should be dominated by scientific norms. Such fora should receive funding independent of any single participant, should be long term (at least one year), and should have a balance of perspectives represented.

• To assist in the process of defining science and management goals, professional scientific associations, groups of scientists, and university research consortia are encouraged to develop syntheses of the state of knowledge and develop plans for strategic research on important coastal problems.

These efforts should be guided by information about research priorities provided by policymakers. The scientific community could help improve the application of appropriate scientific information to coastal management problems by developing consensus-forming processes that support credible analyses for use in policymaking. Many descriptions of priority environmental quality issues have been developed. A good recent summary is given by NRC (1994a).

• Congress and state legislatures should amend legislation to remove barriers to the exchange of information between state and federal levels and between governmental agencies and external scientists, while preserving the intent of such legislation (NRC, 1995c).

Enhance Communications Among Scientists, Policymakers, and the Public

• Policymakers and implementers are encouraged to clearly identify their short- and long-term research needs, and to indicate how the information is to be used, what resources are available to support the collection and analysis of information about natural and social systems, and when the information is needed (NRC, 1995b).

Lists of the priority scientific activities should be developed by individual agencies, or as cooperative efforts among them, in collaboration with scientists and stakeholders. Such lists could form the basis for requests for proposals issued for applied research (NRC, 1995b). New means should be developed to communicate this vital information to scientists and to provide incentives to encourage scientists to carry out identified research. These mechanisms should be designed to improve the transfer of information to scientists on a regular basis (biannually to monthly) as well as on an immediate basis for urgent situations. Such a process could be enhanced by forming a network of science and management professionals interested in cooperation and the timely exchange of information. Improved communications and increased interactions among scientists, policymakers, and the public could make political processes more fact-based and predictable.

• Federal, state, and local entities, with the assistance of universities, nongovernmental organizations (NGOs), and others, should ensure that the results of policy-relevant scientific research are summarized in a manner intelligible to the lay public and are widely disseminated to decisionmakers and the public through various media, including electronic information networks (NRC, 1995c).

It was evident throughout the regional symposia that the public is often a missing component in the application of science to coastal policy. Although an informed public will not always agree with scientists and managers, their reasons for disagreement will more likely be based on knowledge, allowing the possibility of informed compromise. Government agencies and the scientific community must, on a continuing basis, take actions to increase public understanding and awareness of the relative roles of science and policy and the importance to policymakers and implementers of objective, credible, and timely scientific information. These should include communication by scientists to the public and policymakers about the role of science and its limitations (NRC, 1995c).

• Federal, state, and local entities, scientists, NGOs, and others should assist representatives of the print, radio, and television media understand and disseminate the results of policy-relevant scientific research (NRC, 1995b).

Special awards for science and environmental reporting could improve the quality of media coverage. For example, the American Geophysical Union (a professional society) recognizes high-quality science reporting on geoscience issues through its Walter Sullivan Award.

Build Capacity for Science-Policy Interactions

• Federal, state, and local entities that have made innovative efforts to apply scientific expertise in the design and implementation of coastal programs (e.g., EPA's National Estuary Program and the Chesapeake Bay and Great Lakes programs) should be encouraged to prepare assessments of effective models for science-policy interaction that can be used as a guide for implementation in other relevant contexts.

• Federal, state, and local entities should encourage staff scientists to maintain their expertise and stay current with developments in knowledge and technology in their fields (NRC, 1995a).

• Institutions of higher education, as well as individual scientists, should be encouraged to:

— Improve the cross-disciplinary training of natural and social scientists—for example, by enhancing existing programs of advanced training in the marine-oriented natural sciences by including additional training in the social sciences and policy (to attain policy literacy); by enhancing existing programs of advanced training in the marine-oriented social sciences to include additional training in the natural sciences (to attain natural science literacy); and by enhancing or creating programs of training for "science translators" (NRC, 1994a, 1995b,c).

Training programs for science translators should include exposure to the natural and social sciences, policy development and implementation, and conflict management and communication skills. Science translators should not substitute for the involvement of scientists and policymakers directly with one another. Translators can provide a supplementary means to draw practitioners from the two fields together and to help them communicate more effectively with one another and with the public. Scientists and policymakers need to understand each other to work together in defining coastal environmental problems, by posing the appropriate research questions, explaining methods and results, and exploring the possible implications and policy responses to the research results (Douglas, 1995).

— Create consortia for strategic research, in collaboration with federal, state, and local authorities. The consortia should facilitate regular communication of state-of-the-art science to policymakers. This could be accomplished in week-long summer "institutes," individual seminars, and trips to research sites or laboratories (Glidden, 1995). In addition, consortia could sponsor summer internships for graduate students and faculty to work in policymaking organizations (NRC, 1995c). Such consortia could sponsor the joint preparation of written plans describing how science and policy will be integrated in coastal management programs.

— Modify the academic reward system to encourage the involvement of scientists in the policy development and implementation process (NRC, 1995b) by recognizing scholarship in synthesis and application as well as discovery and teaching (Boyer, 1990). Although this is a daunting task, the lack of incentives to encourage involvement of academic scientists in coastal management problems was often cited in the regional symposia as a major barrier to effective use of science in policymaking.

— Encourage the application of scientific knowledge in the development of coastal policies by working closely with state coastal zone management programs and other state agencies. This may require new legislation at the federal and state levels. Participants in the Gulf of Maine symposium believed that state-level environmental impact assessment processes, akin to the federal process required under the National Environmental Policy Act, should be developed (NRC, 1995b).

Management programs should be evaluated, in part, relative to their efforts to and successes in incorporating science in their activities. Research and management reviews of coastal environmental management programs, such as the Section 312 reviews of state coastal management programs, and the activities of the National Estuarine Research Reserves, National Marine Sanctuaries, and the National Estuary Program, should be coordinated or integrated (NRC, 1995b).

ISSUE 2
INTEGRATED AND ADAPTIVE MANAGEMENT

Finding: Coastal policies tend to lack sufficient flexibility and are most often designed to manage single issues.

Often, coastal management is conducted separately by different levels of government (e.g., state versus federal agencies) and by different agencies in state governments or the federal government. Management usually is not integrated or coordinated among entities in a meaningful way to encompass all relevant aspects of a given coastal environmental issue. Actions taken by different parts of government often conflict owing to such factors as divergent legislative mandates, agency cultures, lack of communication, and constituency pressures. Lack of coordination may relate to different groups within a single political jurisdiction but are even more challenging when more than one political jurisdiction is involved.

Adaptive environmental management (see pp. 61-62) implies regular evaluation of management success as measured by some predetermined variables and predictive scientific approaches that can be used to assess and predict risk and to estimate uncertainties related to management processes. There are few instances of adaptive management being used formally as a regular part of coastal management programs. In some cases, adaptive management plans have been developed, but have not been implemented (e.g., coastal zone management and national estuary programs) due to lack of funding.

Recommendation: Employ integrated and adaptive management approaches in coastal policymaking and implementation.

Policy and management processes should be integrated, so that all sectors, political and administrative jurisdictions, stakeholders, and scientific disciplines relevant to particular coastal issues or problems, are included in the process. This should include linkages between land use and marine environmental quality (Terkla, 1995). Policy and management processes should be adaptive in that new data and information, analytical and evaluative techniques, and lessons from management experience are continually incorporated into the process (NRC, 1995b). A clear set of quantitative goals should be articulated for coastal management programs to achieve in a prescribed period of time. For example, coastal management programs might select goals related to wetlands protection, beach and dune management, public access, management of coastal development to reduce losses from natural hazards, and nonpoint-source pollution (Knecht, 1995). Greater specificity and accountability should be built into coastal management systems (Knecht, 1995), with emphasis on outcome-oriented goals. One means of improving linkages would be a greater use of economic methods for valuing, prioritizing, and allocating scarce resources for monitoring (Terkla, 1995). Traditional and nontraditional (Odum, 1995) valuation methods should be used for the allocation of natural resources such as fish, fresh water, and habitat. Specific recommendations include the following:

• Federal, state, and local entities should strive to work across agency boundaries to develop integrated management programs that are adaptive in their formulation and implementation.

• Government entities that have developed programs to achieve integrated and adaptive management (e.g., coastal zone management programs, NEPs, other interagency ecosystem management efforts) should allocate sufficient resources to implement such programs.

Participants in the Gulf of Maine symposium suggested that cumulative impact problems could be reduced if coastal management agencies develop area-wide comprehensive planning programs for all sectors of the coast (e.g., Comprehensive Conservation and Management Plans of the NEPs or Areas of Critical Environmental Concern in Massachusetts) (NRC, 1995b). Rational schemes to manage cumulative impacts should include management goals that are conceptually clear, that demonstrate causal relationships and infrastructure to allow the calculation of key thresholds and monitoring of conditions, and that have adequate capacity for governance (NRC, 1995a). Governance must be flexible so that it can be adapted to greater or lesser intervals of time and geographic area as more information is gathered regarding a coastal environmental issue.

• State legislatures are encouraged to evaluate the performance of their coastal programs by requiring the application of scientific expertise to such evaluations.

• Concerted efforts should be made to assess changes in conditions of coastal environments, resources, and human populations and the degree of achievement of policy goals as a key requirement for adaptive management.

These should include strategic assessments and monitoring programs of a national scope, yielding a periodic (e.g., every five years) assessment of the "state of the coast." However, these efforts should ensure that the information provided covers appropriate spatial and temporal scales to be useful for coastal policy and management at the state and regional levels. Present federal monitoring programs tend to monitor conditions too sparsely over the geographic areas that most concern state and local managers.

• Scientists should improve the application of predictive approaches (see pp. 45-50) to policy development and implementation, including modeling and risk assessment, complete with estimates of their associated uncertainty.

Such information should be used to build integrated multidisciplinary (natural and social sciences) models of systems that need to be understood better. These models should take into consideration, while at the same time striving to overcome, impediments to the effective use of models (NRC, 1995c), including incompleteness of models, imperfect input data, and lack of a widely accepted means to combine environmental and economic factors in a model. Science activities should be focused on making predictions and identifying variables that create uncertainties in these predictions. Modeling should be linked with monitoring and research in an adaptive management framework.

ISSUE 3
ALLOCATION OF RESOURCES

Finding: The allocation of available resources for coastal science and policymaking is suboptimal because few of the resources are devoted to making the connections necessary to promote the appropriate use of science in policymaking.

A great deal of human, fiscal, and physical resources are presently devoted to coastal science and management. For example, the federal government expended a total of $672 million on coastal research in FY1991-1993 (SUSCOS, 1993). U.S. expenditures for management and protection of coastal areas are difficult to estimate but may equal or exceed the research expenditure. Despite these large investments, the use of science in coastal policy development, implementation, and evaluation has not been as effective as desired, in part because little support has been directed to disseminating information between scientists and policymakers, promoting interdisciplinary research teams, integrating science and policy components of individual agencies, and sharing personnel among science and policy portions of agencies. The reallocation of existing resources could draw coastal science and policy into a more cooperative endeavor.

Recommendation: Improve the allocation and coordination of resources to achieve effective interaction between coastal scientists and policymakers.

Although the allocation of some new resources (fiscal, physical, and human) will be needed to increase the use of science in coastal policymaking, much can be accomplished through better mobilization and coordination of existing resources. Additional resources may be needed to build long-term data bases needed to guide management (Glidden, 1995). It was noted at the California symposium (NRC, 1995a) and in a previous NRC assessment (NRC, 1990a) that many resources are wasted on ineffective or unnecessary monitoring. Agencies responsible for coastal ecosystem protection should reevaluate their monitoring priorities and appropriately adjust the focus of their monitoring activities (NRC, 1990b). Many issues could benefit from understanding and documenting past actions and how they affected natural and social systems (NRC, 1995c). Geographic information systems can be used to organize this information (Chang et al., 1995). New monitoring technologies and data collection by community-based volunteers should be explored (Chang et al., 1995). Specific recommendations include the following:

• Federal, state, and local entities should require that a given portion of scientific research budgets be devoted to the translation and dissemination of scientific results.
• Federal, state, and local entities, in their request for proposals, should promote the formation of interdisciplinary teams to carry out policy-relevant research.
• Federal, state, and local entities should develop mechanisms for better integration of their science and policy capabilities, through such means as data sharing, colocation of facilities, and cooperative programs.
• Federal, state, and local entities should facilitate personnel exchange or staff-sharing arrangements, whereby scientists and NGO and industry personnel may spend time in government, and government employees can work in universities, NGOs, and corporations on temporary assignments.

The committee offers the recommendations in this chapter as suggestions that could be implemented immediately. Undoubtedly, as more experience is gained using science in coastal management, new ideas for further improvements will emerge.

References

Acheson, J.M. 1995. Environmental protection, fisheries management, and the theory of chaos. In *Improving Interactions Between Coastal Science and Policy: Proceedings of the Gulf of Maine Symposium.* National Academy Press, Washington, D.C.

Albermarle-Pamlico Estuarine Study. 1995. *Environmental and Economic Stewardship in the Albermarle-Pamlico Region: A Comprehensive Conservation and Management Plan.* North Carolina Department of Environment, Health, and Natural Resources, Raleigh, N.C.

Anderson, J. 1984. *Public Policy-Making.* Holt, Rinehart and Winston, New York.

Bernstein, B.B., B.E. Thompson, and R.W. Smith. 1991. A combined science and management framework for developing regional monitoring objectives. *Coastal Management* 21:185-195.

Boesch, D., and S-A. Macke. 1995. Bridging the gap: what natural scientists and policymakers need to know about each other. Pp. 33-48 in *Improving Interactions Between Coastal Science and Policy: Proceedings of the California Symposium.* National Academy Press, Washington, D.C.

Boesch, D.F., and N.N. Rabalais (eds.). 1987. *Long-Term Environmental Effects of Offshore Oil and Gas Development.* Elsevier Applied Science, London.

Bower, B.T. 1992. *Producing Information for Integrated Coastal Management Decisions.* National Oceanic and Atmospheric Administration, Washington, D.C.

Boyer, E.L. 1990. *Scholarship Reconsidered: Priorities of the Professoriate.* Carnegie Foundation for the Advancement of Teaching, Princeton, N.J.

Boynton W.R., J.H. Garber, R. Summers, and W.M. Kemp. 1995. Inputs, transformations, and transports of nitrogen and phosphorus in Chesapeake Bay and selected tributaries. *Estuaries* 18:285-314.

Cairncross, F. 1991. *Costing the Earth.* Harvard Business School Press, Boston.

Caldwell, L. 1990. *Between Two Worlds: Science, the Environmental Movement, and Policy Choice.* Cambridge University Press, Cambridge.

California Coastal Commission. 1987. *California Coastal Resources Guide.* University of California Press, Berkeley.

California Coastal Commission. 1994. *Regional Cumulative Assessment Project: Preliminary Findings and Recommendations—Monterey Bay Region.* San Francisco.

California Coastal Zone Conservation Commission. 1975. *California Coastal Plan.* p. 39.

Carlton, J.T., and J.B. Geller. 1993. Ecological roulette, the global transport of nonindigenous marine organisms. *Science* 261:78-82.

Chang, B.D., R.L. Stevenson, D.J. Wildish, and W.M. Watson-Wright. 1995. Protecting regionally significant marine habitats in the Gulf of Maine: A Canadian perspective. In *Improving Interactions Between Coastal Science and Policy: Proceedings of the Gulf of Maine Symposium.* National Academy Press, Washington, D.C.

Chua, T.-E. 1993. Essential elements of integrated coastal management. *Ocean and Coastal Management* 21:81-108.

Connor, M.S. 1995. The Boston Harbor case: management and science. In *Improving Interactions Between Coastal Science and Policy: Proceedings of the Gulf of Maine Symposium.* National Academy Press, Washington, D.C.

Culliton, T.J., M.A. Warren, T.R. Goodspeed, D.G. Remer, C.M. Blackwell, and J. MacDonough. 1990. *Fifty Years of Population Change Along the Nation's Coast.* Second Report of the Coastal Trends Series. Office of Oceanography and Marine Assessment, National Oceanic and Atmospheric Administration, Washington, D.C.

Dickert, T.G., and A.E. Tuttle. 1985. Cumulative impact assessment in environmental planning: A coastal wetland watershed example. *Environmental Impact Assessment Review* 5:37-64.

Douglas, P.M. 1995. What do policymakers and policy-implementors need from scientists? Pp. 15-32 in *Improving Interactions Between Coastal Science and Policy: Proceedings of the California Symposium.* National Academy Press, Washington, D.C.

Earll, R.C. 1992. Common sense and the precautionary principle: an environmentalist's perspective. *Marine Pollution Bulletin* 24:182-186.

Ebbesmeyer, C.C., and W.J. Ingraham, Jr. 1992. Shoe spill in the North Pacific. *EOS, Transactions of the American Geophysical Union* 73:361,365.

Environmental Protection Agency (EPA). 1992. *Framework for Ecological Risk Assessment.* EPA/630/R-92/001. EPA, Washington, D.C.

Environmental Protection Agency (EPA) Science Advisory Board. 1990. *Reducing Risk: Setting Priorities and Strategies for Environmental Protection.* EPA, Washington, D.C.

Fogerty, M.J., M.P. Sissenwine, and E.B. Cohen. 1991. Recruitment variability and the dynamics of exploited marine populations. *Trends in Ecology and Evolution* 6:241-246.

Fortman, L. 1990. The role of professional norms and beliefs in the agency-client relations of natural science bureaucracies. *Natural Resources Journal* 30(3):361-380.

Freudenberg, W.R., and R. Gramling. 1994. *Oil in Troubled Waters: Perceptions, Politics, and the Battle Over Offshore Drilling.* State University of New York Press, Albany.

Glidden, T. 1995. Cumulative impacts in Gulf of Maine estuaries: policy development considerations. In *Improving Interactions Between Coastal Science and Policy: Proceedings of the Gulf of Maine Symposium.* National Academy Press, Washington, D.C.

Goldberg, E.D. 1986. TBT: an environmental dilemma. *Environment* 28:17-22.

Goldberg, E.D. 1992. Marine metal pollutants. A small set. *Marine Pollution Bulletin* 25:1-4.

Gray, J.S. 1990. Statistics and the precautionary principle. *Marine Pollution Bulletin* 21:174-176.

Gray, J.S., D. Calamari, R. Duce, J.E. Protmann, P.G. Wells, and H.L. Windom. 1991. Scientifically based strategies for marine environmental protection and management. *Marine Pollution Bulletin* 22:432-440.

Group of Experts on the Scientific Aspects of Marine Pollution (GESAMP). 1990. *The State of the Marine Environment.* Blackwell Scientific Publications, Oxford.

Hallegraff, G.M., and C.J. Bolch. 1991. Transport of toxic dinoflagellate cysts via ship ballast water. *Marine Pollution Bulletin* 22:27-30.

Hammond, K., and L. Adelman. 1976. Science, values and human judgment. *Science* 194:389-396.

Harris, M. 1968. *The Rise of Anthropological Theory: A History of the Theories of Culture.* Cromwell, New York.

Hedgpeth, J.W. 1993. Foreign invaders. *Science* 261:34-35.

International Association for Great Lakes Research (IAGLR). 1991. *The 34th Conference of the International Association for Great Lakes Research, Program and Abstracts.* University of Michigan, Ann Arbor.

Jasanoff, S. 1990. *The Fifth Branch: Science Advisors as Policy-Makers.* Harvard University Press, Cambridge, Mass.

Jaworski, N.S. 1990. Retrospective of the water quality of the upper Potomac estuary. *Aquatic Science* 3:11-40.

Knecht, R.W. 1995. On the role of science in the implementation of national coastal management programs. In *Improving Interactions Between Coastal Science and Policy: Proceedings of the Gulf of Maine Symposium.* National Academy Press, Washington, D.C.

Lampl, L.L. 1995. On the subject of shellfish, water quality, and human groups. In *Improving Interactions Between Coastal Science and Policy: Proceedings of the Gulf of Mexico Symposium.* National Academy Press, Washington, D.C.

Lee, K. 1993. *Compass and Gyroscope. Integrating Science and Politics for the Environment.* Island Press, Washington, D.C.

Louisiana Department of Natural Resources. 1991. *LEAP to 2000: Louisiana Environmental Action Plan.* Department of Natural Resources, Baton Rouge.

McKinney, L. 1995. The role of freshwater inflows in estuarine ecosystems of the Gulf of Mexico. In *Improving Interactions Between Coastal Science and Policy: Proceedings of the Gulf of Mexico Symposium.* National Academy Press, Washington, D.C.

Millsap, W. 1984. *Applied Social Science for Environmental Planning.* Westview Press, Boulder, Colo.

Nader, L. 1969. *Law and Culture in Society.* Aldine Publishers, Chicago.

National Academy of Sciences (NAS). 1957. *The Effects of Atomic Radiation on Oceanography and Fisheries.* NAS, Washington, D.C.

National Marine Fisheries Service (NMFS). 1993. *Our Living Oceans.* National Oceanic and Atmospheric Administration, Department of Commerce, Washington, D.C.

National Ocean Service. 1995. *Healthy Coastal Ecosystems and the Role of Integrated Coastal Management.* National Ocean Service, National Oceanic and Atmospheric Administration, Washington, D.C.

National Oceanic and Atmospheric Administration (NOAA). 1990. *Coastal Environmental Quality in the United States, 1990: Chemical Contamination in Sediments and Tissues.* National Oceanic and Atmospheric Administration, Rockville, Md.

National Research Council (NRC). 1971. *Radioactivity in the Marine Environment.* National Academy Press, Washington, D.C.

National Research Council (NRC). 1978. *OCS Oil and Gas: An Assessment of the Department of Interior Environmental Studies Program.* National Academy Press, Washington, D.C.

National Research Council (NRC). 1983a. *Drilling Discharges in the Marine Environment.* National Academy Press, Washington, D.C.

National Research Council (NRC). 1983b. *Risk Assessment in the Federal Government: Managing the Process.* National Academy Press, Washington, D.C.

National Research Council (NRC). 1989. *The Adequacy of Environmental Information for Outer Continental Shelf Oil and Gas Decisions: Florida and California.* National Academy Press, Washington, D.C.

National Research Council (NRC). 1990a. *Monitoring Southern California's Coastal Waters.* National Academy Press, Washington, D.C.

National Research Council (NRC). 1990b. *Managing Troubled Waters: The Role of Marine Environmental Monitoring.* National Academy Press, Washington, D.C.

National Research Council (NRC). 1990c. *Assessment of the U.S. Outer Continental Shelf Environmental Studies Program.* National Academy Press, Washington, D.C.

National Research Council (NRC). 1991. *The Adequacy of Environmental Information for Outer Continental Shelf Oil and Gas Decisions: Georges Bank.* National Academy Press, Washington D.C.

National Research Council (NRC). 1992a. *Restoration of Aquatic Ecosystems: Science, Technology, and Public Policy.* National Academy Press, Washington, D.C.

National Research Council (NRC). 1992b. *Oceanography in the Next Decade: Building New Partnerships.* National Academy Press, Washington, D.C.

National Research Council (NRC). 1993. *Managing Wastewater in Coastal Urban Areas.* National Academy Press, Washington, D.C.

National Research Council (NRC). 1994a. *Priorities for Coastal Ecosystem Science.* National Academy Press, Washington, D.C.

National Research Council (NRC). 1994b. *Environmental Science in the Coastal Zone: Issues for Further Research.* National Academy Press, Washington, D.C.

National Research Council (NRC). 1995a. *Improving Interactions Between Coastal Science and Policy: Proceedings of the California Symposium.* National Academy Press, Washington, D.C.

National Research Council (NRC). 1995b. *Improving Interactions Between Coastal Science and Policy: Proceedings of the Gulf of Maine Symposium.* National Academy Press, Washington, D.C.

National Research Council (NRC). 1995c. *Improving Interactions Between Coastal Science and Policy: Proceedings of the Gulf of Mexico Symposium.* National Academy Press, Washington, D.C.

Nichols, F.H., J.K. Thompson, and L.E. Schemel. 1990. Remarkable invasion of San Francisco Bay (California, USA) by the Asian clam, *Potamocorbula amurensis*: displacement of a former community. *Marine Ecology Progress Series* 66:95-101.

North Sea Interministerial Conference. 1900. *Final Declaration of the Third International Conference on the Protection of the North Sea.* The Hague, The Netherlands.

Odum, H.T. 1995. Economic impacts brought about by alterations to freshwater flow. In *Improving Interactions Between Coastal Science and Policy: Proceedings of the Gulf of Mexico Symposium.* National Academy Press, Washington, D.C.

OECD. 1991. *Report on Coastal Zone Management: Integrated Policies and Draft Recommendations of the Council on Integrated Coastal Management.* Paris.

Officer, C.B., R.B. Biggs, J.L. Taft, L.E. Cronin, M.A. Tyler, and W.R. Boynton. 1984. Chesapeake Bay anoxia: Origin, development, and significance. *Science* 223:22-27.

Orbach, M. 1995. Social scientific contributions to coastal policy making. Pp. 49-59 in *Improving Interactions Between Coastal Science and Policy: Proceedings of the California Symposium.* National Academy Press, Washington, D.C.

Owens, M. and J.C. Cornwell. 1995. Sedimentary evidence for decreased heavy-metal inputs to the Chesapeake Bay. *Ambio* 24(1):24-27.

Parker, C.A., and J. O'Reilly. 1991. Oxygen depletion in Long Island Sound: A historical perspective. *Estuaries* 14:248-264.

Peterson, J. 1984. *Citizen Participation in Science Policy.* University of Massachusetts Press, Amherst.

Rabalais, N.N., R.E. Turner, and W.J. Wiseman, Jr. 1994. Hypoxic conditions in bottom waters on the Louisiana-Texas shelf. Pp. 50-54 in M.J. Dowgiallo (ed.), *Coastal Oceanographic Effects of 1993 Mississippi River Flooding.* Special NOAA Report. NOAA Coastal Ocean Program Office/National Weather Service, Silver Spring, Md.

Rabalais, N.N., Q. Dortch, D. Justic, M.B. Kilgen, P.H. Templet, R.E. Turner, B. Cole, D. Duet, M. Beacham, S. Lentz, M. Parsons, S. Rabalais, and R. Robichaux. 1995. *Characterization of the Current Status and Historical Trends of Eutrophication, Pathogen Contamination, and Toxic Substances in the Barataria and Terrebonne Estuarine Systems.* BTNEP Publication, Barataria-Terrebonne National Estuary Program, Thibodaux, La.

Richkus, W.A., H.M. Austin, and S.J. Nelson. 1992. Fisheries assessment and management synthesis: lessons for Chesapeake Bay. Pp. 75-114 in *Perspectives on Chesapeake Bay, 1992: Advances in Estuarine Sciences.* Scientific and Technical Advisory Committee, Chesapeake Bay Program, Solomons, Md.

Rydberg, L., L. Elder, S. Floderus, and W. Graneli. 1990, Interaction between supply of nutrients, primary production, sedimentation, and oxygen consumption in SE Kattegat. *Ambio* 19:134-141.

Sabatier, P.A. 1995. Alternative models of the role of science in public policy: applications to coastal zone management. Pp. 83-95 in *Improving Interactions Between Coastal Science and Policy: Proceedings of the California Symposium.* National Academy Press, Washington, D.C.

Scheiber, H.N. 1995. Success and failure in science-policy interactions: cases from the history of California coastal and ocean studies, 1945-1973. Pp. 97-122 in *Improving Interactions Between Coastal Science and Policy: Proceedings of the California Symposium.* National Academy Press, Washington, D.C.

Shelley, P., and E. Dorsey. 1995. Policy development considerations with regionally significant habitats: science and policy in federal fisheries management. In *Improving Interactions Between Coastal Science and Policy: Proceedings of the Gulf of Maine Symposium.* National Academy Press, Washington, D.C.

Sklar, F.H. 1995. Coastal Gulf of Mexico environmental impacts brought about by alterations to freshwater flow. In *Improving Interactions Between Coastal Science and Policy: Proceedings of the Gulf of Mexico Symposium.* National Academy Press, Washington, D.C.

Sorensen, J.C., and S.T. McCreary. 1990. *Institutional Arrangements for Managing Coastal Resources and Environments.* National Park Service, U.S. Department of the Interior, Washington, D.C.

Subcommittee on U.S. Coastal Ocean Science (SUSCOS). 1993. *Setting a New Course for U.S. Coastal Ocean Science. Phase I: Inventory of Federal Programs.* Federal Coordinating Council on Science, Engineering, and Technology, Washington, D.C.

Terkla, D.G. 1995. Economic issues in monitoring marine water quality. In *Improving Interactions Between Coastal Science and Policy: Proceedings of the Gulf of Maine Symposium.* National Academy Press, Washington, D.C.

The Year 2020 Panel. 1988. *Population Growth and Development in the Chesapeake Bay Watershed to the Year 2020.* Chesapeake Executive Council, Annapolis, Md.

United Nations Conference on Environment and Development (UNCED). 1992b. Agenda 21, Chapter 17 *"Protection of the oceans, all kinds of seas, including enclosed and semi-enclosed seas, and coastal areas and protection, rational use and development of their living resources."* New York and Geneva.

Van der Weide, J. 1993. A systems view of integrated coastal management. *Oceans and Coastal Management* 21:129-148.

Vermont Agency for Natural Resources. 1991. *Environment 1991: Risks to Vermont and Vermonters.* Vermont Agency for Natural Resources, Waterbury.

Weiss, C. (ed.). 1987. *Using Social Research in Public Policy-Making.* Lexington Books, Lexington, Mass.

Wheelwright, J. 1994. *Degrees of Disaster: Prince William Sound: How Nature Reels and Rebounds.* Simon & Schuster, New York.

Appendix

Biographies of Committee Members

Donald F. Boesch is president of the University of Maryland Center for Environmental and Estuarine Studies. He is also a professor of marine science at the center. Before joining the University of Maryland, Dr. Boesch was for 10 years a professor of marine science at Louisiana State University and the first executive director of the Louisiana Universities Marine Consortium. He earned a Ph.D. from the College of William and Mary. Dr. Boesch's research interests include benthic ecology, coastal wetlands, and the interdisciplinary science of estuarine and continental shelf environments. He has been very involved in national and regional environmental science and policy issues and has served on numerous federal advisory committees and National Research Council boards and committees.

Biliana Cicin-Sain is presently a professor of marine studies in the Graduate College of Marine Studies at the University of Delaware, where she also holds joint appointments in the Department of Political Science and the College of Urban Affairs and Public Policy. Dr. Cicin-Sain serves as codirector of the Center for the Study of Marine Policy at the University of Delaware and as editor-in-chief of *Ocean and Coastal Management,* an international journal devoted to the analysis of all aspects of ocean and coastal management. Dr. Cicin-Sain has written extensively on a range of marine policy issues, including fisheries management, marine mammal management, offshore oil development, multiple-use conflicts, and international marine policy. In the past several years, her work has emphasized issues related to the achievement of integrated ocean and coastal management policies.

Peter M. Douglas has been employed by the California Coastal Commission since 1977 and has been its executive director since 1985. He earned a J.D. from the University of California, Los Angeles, in 1969. Mr. Douglas is responsible for policymaking and implementation of a comprehensive coastal and ocean resource management program pursuant to state and federal law. His interests include building more effective bridges between the scientific and public policy decisionmaking communities.

Edward D. Goldberg has been associated with the Scripps Institution of Oceanography since 1949. He was appointed professor of chemistry there in 1961. Dr. Goldberg earned a Ph.D. in chemistry from the University of Chicago in 1949. His scientific interests include the geochemistry of natural waters and sediments, the demography of the coastal zone, the history of burning, waste management, and marine pollution. Dr. Goldberg currently serves as editor of a technical series in oceanography, *The Sea: Ideas and Observations*, and was a coeditor of two volumes, *Earth Sciences and Meteorites* and *Man's Impact on Terrestrial and Marine Ecosystems*. Dr. Goldberg is a member of the National Academy of Sciences.

Susan S. Hanna is associate professor of marine economics in the Department of Agricultural and Resource Economics at Oregon State University. She also directs the research program on Property Rights and the Performance of Natural Resource Systems at the Beijer Institute, the Royal Swedish Academy of Sciences, Stockholm, Sweden. Dr. Hanna conducts research in the areas of economics of fisheries, fishery management and regulation, ocean-use interactions, seafood markets and fishery regulation, user participation in resource management, institutional economics, property rights and marine resources, and economic history. Dr. Hanna has performed a variety of scientific/policy outreach activities as a member of advisory committees and is presently a member of the National Research Council's Ocean Studies Board.

David H. Keeley has 15 years of experience in environmental management, policy development, and planning, with an emphasis on coastal and estuarine issues. He has worked at the local, county, and state levels in a variety of land-use planning roles. Mr. Keeley was instrumental in forming the Gulf of Maine Program, a state-provincial environmental and economic initiative. He has managed more than $30 million in grants and supervises a staff of planners, lawyers, and scientists. Mr. Keeley is active at the national level, recently completing a two-year term as chairman of the Coastal States Organization. He presently serves on numerous state, national, and international advisory panels and boards.

Michael K. Orbach is a professor of marine affairs and policy at Duke University. From 1983 to 1993 he was a professor of anthropology at East Carolina

University. From 1976 to 1979 he served as social anthropologist and social science advisor to the National Oceanic and Atmospheric Administration in Washington, D.C. He has published widely on marine social science topics, including fisheries limited entry and effort management, Indochinese fishermen adaptation, marine mammal-fishery interactions, and state, regional, and federal fisheries and marine policy, including *Hunters, Seamen and Entrepreneurs,* an ethnography of the San Diego tuna fishermen.

John M. Teal has been a senior scientist at the Woods Hole Oceanographic Institution since 1971. He earned a Ph.D. from Harvard University in 1955. Dr. Teal's primary specialty is coastal wetlands ecology, but he has worked on ecology and physiology in a variety of coastal and oceanic systems. His wetlands research has, for the past 20 years, included much work on coastal pollution and its effects on coastal ecosystems, including but not limited to wetlands. The pollutants of interest have included oil, heavy metals, and especially nutrients responsible for coastal eutrophication. He has served on a variety of local to national committees concerned with pollution, its remediation, and interactions between scientists and managers or scientists and the public on these issues. He has authored or coauthored some 140 professional papers and four books for the general public.